Micro/Nano Manufacturing

Special Issue Editors

Hans Nørgaard Hansen
Guido Tosello

MDPI • Basel • Beijing • Wuhan • Barcelona • Belgrade

MDPI

Special Issue Editors

Hans Nørgaard Hansen
Technical University of Denmark
Denmark

Guido Tosello
Technical University of Denmark
Denmark

Editorial Office
MDPI AG
St. Alban-Anlage 66
Basel, Switzerland

This edition is a reprint of the Special Issue published online in the open access journal *Micromachines* (ISSN 2072-666X) in 2017 (available at: http://www.mdpi.com/journal/micromachines/special_issues/micro_nano_manufacturing).

For citation purposes, cite each article independently as indicated on the article page online and as indicated below:

Author 1; Author 2. Article title. *Journal Name* **Year**, *Article number*, page range.

First Edition 2017

ISBN 978-3-03842-604-2 (Pbk)
ISBN 978-3-03842-605-9 (PDF)

Table of Contents

About the Special Issue Editors

Hans Nørgaard Hansen is working in the field of Micro/Nano Manufacturing. He deals with establishing the basis of industrial production of products and components in metal, polymers and ceramics in the micro- and nanometer regime. Specific activities include product development, materials development and development of processes and production systems focused on micro mechanical systems. The core activities comprise micro product design and development, tooling technologies for micro injection moulding and micro metal forming, mass production technologies, chemical and electrochemical processes and laser technologies. Finally the integration of processes into coherent process chains is a key activity. Professor Hans Nørgaard Hansen is the Head of the Department of Mechanical Engineering at the Technical University of Denmark, is a Fellow of CIRP (International Academy of Production Engineering) and the 2015–2017 President of euspen (European Society of Precision Engineering and Nanotechnology).

Guido Tosello is Associate Professor at the Technical University of Denmark, Department of Mechanical Engineering, Section of Manufacturing Engineering. Guido's principal research interests are the analysis, characterization, monitoring, control, optimization and simulation of precision moulding processes at micro/nano scales of thermoplastic materials. Technologies supporting precision/micro/nano moulding processes are of research interest: advanced process chain for micro/nano tools manufacturing, precision and micro additive manufacturing, dimensional and surface micro/nano metrology.

Guido Tosello is the recipient of the "Technical University of Denmark Best PhD Research Work 2008 Prize" for his PhD thesis "Precision Moulding of Polymer Micro Components", of the 2012 Alan Glanvill Award by The Institute of Materials, Minerals and Mining (IOM3) (UK), given as recognition for research of merit in the field of polymeric materials, of the Young Research Award 2014 from the Polymer Processing Society (USA) in recognition of scientific achievements and research excellence in polymer processing within 6 years from PhD graduation, and of the Outstanding Reviewer Award 2016 of the Institute of Physics (UK) for his contribution to the Journal of Microengineering and Micromechanics.

Guido Tosello is currently the Project Coordinator of the Horizon2020 European Marie Skłodowska-Curie Innovative Training Network MICROMAN "Process Fingerprint for Zero-defect Net-shape MICROMANufacturing" (2015–2019).

Preface to "Micro/Nano Manufacturing"

Micro- and nano-scale manufacturing has been the subject of an increasing amount of interest and research efforts worldwide in both academia and industry over the past 10 years. Traditional lithography-based technology forms the basis of silicon-based micro-electro-mechanical systems (MEMS) manufacturing, but also precision manufacturing technologies have been developed to cover micro-scale dimensions and accuracies. Furthermore, these fundamentally different technology ecosystems are currently combined in order to exploit the strengths of both platforms. One example is the use of lithography-based technologies to establish nanostructures that are subsequently transferred to 3D geometries via injection molding.

Manufacturing processes at the micro-scale are the key-enabling technologies to bridge the gap between the nano- and the macro-worlds, to increase the accuracy of micro/nano-precision production technologies, and to integrate different dimensional scales in mass-manufacturing processes. Accordingly, the present Special Issue provides recent developments in the field of micro/nano manufacturing in terms of production techniques and key enabling technologies that push the boundaries of the state of the art mass-manufacturing of micro-scale and micro/nano structured components.

The Special Issue consists of 13 original research papers, which cover both fundamental process technology developments as well as the application of those technologies for the fabrication of micro/nano devices. The papers included in this Special Issue address research in four main domains of micro/nano manufacturing:

(1) Process developments of micro-scale fabrication methods. Cannella et al. [1] developed a novel tooling system for in-die sintering of micro metal gears; Martinez-López et al. [2] investigated xurography and laser ablation processes for the rapid fabrication of micromixing arrays; Obikawa et al. [3] demonstrated ultrasonic-assisted incremental microforming for the rapid prototyping of thin shell metal pyramids.

(2) Micro/nano manufacturing technologies based on electrochemical processing. Bahr et al. [4] investigated the slurry application and injection system to advance the performance of the Chemical Mechanical Planarization (CMP) process; Blondiaux et al. [5] realized the fabrication of nanostructured steel mold inserts by applying a combination of nanosphere lithography and electrochemical etching; Guo et al. [6] demonstrated the fabrication of mesoscale channels by applying a micro electrochemical machining (μECM) process based on the scanning micro electrochemical flow cell (SMEFC).

(3) Quality control of micro systems and processes. Baruffi et al. [7] demonstrated the application of the replica molding technology for quality control of micro milled surfaces; Cao et al. [8] investigated the effect of profile errors on the surface of micro lenses on laser beam homogenization; Choi et al. [9] presented a novel testing platform to characterize and predict the mechanical damage of miniaturized haptic actuator; Gao et al. [10] modeled the effect of micro manufacturing process variations on the characteristics of a micro machined doubly-clamped beam.

(4) Key enabling technologies for micro production. Aizawa et al. [11] presented the development of low-temperature plasma nitriding for the treatment of stainless miniaturized pipes and nozzles' internal surfaces; Davoudinejad et al. [12] modeled and simulated the micro end-milling process including the effect of tool run-out; Wilhelmi et al. [13] demonstrated the improvements in multi-stage micro production by handling wire-based linked micro parts.

We wish to thank all authors who submitted their papers to the Special Issue "Micro/Nano Manufacturing". We would also like to acknowledge all the reviewers whose careful and timely reviews ensured the quality of this Special Issue.

References

1. Cannella, E.; Nielsen, E.; Stolfi, A. Designing a Tool System for Lowering Friction during the Ejection of In-Die Sintered Micro Gears. *Micromachines* **2017**, *8*, 214; doi:10.3390/mi8070214.
2. Martínez-López, J.; Betancourt, H.; García-López, E.; Rodriguez, C.; Siller, H. Rapid Fabrication of Disposable Micromixing Arrays Using Xurography and Laser Ablation. *Micromachines* **2017**, *8*, 144; doi:10.3390/mi8050144.
3. Obikawa, T.; Hayashi, M. Ultrasonic-Assisted Incremental Microforming of Thin Shell Pyramids of Metallic Foil. *Micromachines* **2017**, *8*, 142; doi:10.3390/mi8050142.
4. Bahr, M.; Sampurno, Y.; Han, R.; Philipossian, A. Slurry Injection Schemes on the Extent of Slurry Mixing and Availability during Chemical Mechanical Planarization. *Micromachines* **2017**, *8*, 170; doi:10.3390/mi8060170.
5. Blondiaux, N.; Pugin, R.; Andreatta, G.; Tenchine, L.; Dessors, S.; Chauvy, P.; Diserens, M.; Vuillermoz, P. Fabrication of Functional Plastic Parts Using Nanostructured Steel Mold Inserts. *Micromachines* **2017**, *8*, 179; doi:10.3390/mi8060179.
6. Guo, C.; Qian, J.; Reynaerts, D. Fabrication of Mesoscale Channel by Scanning Micro Electrochemical Flow Cell (SMEFC). *Micromachines* **2017**, *8*, 143; doi:10.3390/mi8050143.
7. Baruffi, F.; Parenti, P.; Cacciatore, F.; Annoni, M.; Tosello, G. On the Application of Replica Molding Technology for the Indirect Measurement of Surface and Geometry of Micromilled Components. *Micromachines* **2017**, *8*, 195; doi:10.3390/mi8060195.
8. Cao, A.; Pang, H.; Wang, J.; Zhang, M.; Chen, J.; Shi, L.; Deng, Q.; Hu, S. The Effects of Profile Errors of Microlens Surfaces on Laser Beam Homogenization. *Micromachines* **2017**, *8*, 50; doi:10.3390/mi8020050.
9. Choi, B.; Kwon, J.; Jeon, Y.; Lee, M. Development of Novel Platform to Predict the Mechanical Damage of a Miniature Mobile Haptic Actuator. *Micromachines* **2017**, *8*, 156; doi:10.3390/mi8050156.
10. Gao, L.; Zhou, Z.; Huang, Q. Modeling of the Effect of Process Variations on a Micromachined Doubly-Clamped Beam. *Micromachines* **2017**, *8*, 81; doi:10.3390/mi8030081.
11. Aizawa, T.; Wasa, K. Low Temperature Plasma Nitriding of Inner Surfaces in Stainless Steel Mini-/Micro-Pipes and Nozzles. *Micromachines* **2017**, *8*, 157; doi:10.3390/mi8050157.
12. Davoudinejad, A.; Tosello, G.; Parenti, P.; Annoni, M. 3D Finite Element Simulation of Micro End-Milling by Considering the Effect of Tool Run-Out. *Micromachines* **2017**, *8*, 187; doi:10.3390/mi8060187.
13. Wilhelmi, P.; Schenck, C.; Kuhfuss, B. Handling in the Production of Wire-Based Linked Micro Parts. *Micromachines* **2017**, *8*, 169; doi:10.3390/mi8060169.

Hans Nørgaard Hansen and Guido Tosello
Special Issue Editors

micromachines

MDPI

Article

Designing a Tool System for Lowering Friction during the Ejection of In-Die Sintered Micro Gears

Emanuele Cannella [1,2,*] , Emil Krabbe Nielsen [3] and Alessandro Stolfi [2]

[1]   IPU Product Development, Kgs. Lyngby 2800 DK, Denmark
[2]   DTU Department of Mechanical Engineering, Kgs. Lyngby 2800 DK, Denmark; alesto@mek.dtu.dk
[3]   DTU Department of Electrical Engineering, Kgs. Lyngby 2800 DK, Denmark; ekrani@elektro.dtu.dk
*   Correspondence: emcann@ipu.dk; Tel.: +45-4525-6286

Received: 29 May 2017; Accepted: 3 July 2017; Published: 6 July 2017

Abstract: The continuous improvements in micro-forging technologies generally involve process, material, and tool design. The field assisted sintering technique (FAST) is a process that makes possible the manufacture of near-net-shape components in a closed-die setup. However, the final part quality is affected by the influence of friction during the ejection phase, caused by radial expansion of the compacted and sintered powder. This paper presents the development of a pre-stressed tool system for the manufacture of micro gears made of aluminum. By using the hot isostatic pressing (HIP) sintering process and different combinations of process parameters, the designed tool system was compared to a similar tool system designed without a pre-stressing strategy. The comparison between the two tool systems was based on the ejection force and part fidelity. The ejection force was measured during the tests, while the part fidelity was documented using an optical microscope and computed tomography in order to obtain a multi-scale characterization. The results showed that the use of pre-stress reduced the porosity in the gear by 40% and improved the dimensional fidelity by more than 75% compared to gears produced without pre-stress.

Keywords: micro sintering; field assisted sintering technique; hot isostatic pressing; micro gears; computed tomography; dimensional accuracy; porosity analysis

1. Introduction

Micro-sintering enables the manufacture of micro-scale components with a complex shape and geometry. Near-net-shape components can be easily manufactured and post-removal processes, e.g., grinding, are limited to the final surface polishing. Compared to cutting processes, since the material scrap is reduced, a high saving of material is possible. Generally, the conventional sintering process can be considered as two different phases, i.e., compaction and heating of the powder [1]. The powder is fed into a closed-die and the punches apply the desired pressure to compact the raw material. The compacted sample, called "green compact", is ejected from the die and moved into a temperature-controlled oven. The sintering temperature is a process parameter and is strictly dependent on the material. This value never overcomes the melting temperature of the material. Therefore, the green compact is kept inside the oven for a defined time, namely the "sintering time". In a final phase, the sample is moved out from the oven and cooled down to the room temperature. The whole process makes it possible to enable the densification of the sample by diffusional mechanisms issued from the pressure and high temperature. Since, in the sintering process, the two steps, i.e., heating and compaction, are carried out separately, the manufacturing time per component is extended. Additionally, oxidation of the material may occur and a protective atmosphere is required to prevent this phenomenon.

To decrease the sintering time per component, new sintering processes have been developed. Such processes conduct heating and compaction of the powder simultaneously, thereby reducing manufacturing time and improving the densification of the component [2–4]. Such new sintering processes establish a mechanical-thermal field, which enables a different bonding mechanism between the powder particles [5–7]. Hot isostatic pressing (HIP) and field assisted sintering technology (FAST) [8–10], are the main names given to such sintering processes. The main difference is on the kind of heating principle [11]. In the HIP sintering, an external heating element, e.g., a heating nozzle, provides the sintering of the pressed compact. In FAST sintering, the sintering heat is provided from the electrical current generating Joule heating while flowing into the compactor and/or the die [12,13]. FAST and HIP operate both the compaction and heating operations inside the closed-die setup. Since the compacted and heated powder cannot expand radially in such a setup, both sintering approaches generate a radial pressure at the die/sample interface. The radial pressure is mainly caused by the axial compaction pressure and thermal heating. The theoretical models, studied by Long and Bockstiegel [14,15], allow an estimation of the radial pressure caused from the axial compaction of the powder. The thermal influence on the radial pressure is studied by applying the thermal expansion law. The radial pressure has a direct impact on the quality of the sintered part because it increases the friction at the interface and, thus, the ejection force. Additionally, the "spring-back" effect may further affect the quality of the sintered component [16]. Several studies have investigated new solutions to reducing the friction, leading to four different types of approaches. A first approach relies on lubricants, which are commonly applied in conventional sintering. The majority of lubricants evaporate above a temperature of 200 °C, leaving, e.g., graphite or molybdenum disulfide as the only options. However, the use of lubricants influences the sample quality in terms of contamination, reducing the "green strength" and density [17–20]. A second approach is based on the use of a split die solution, where the die includes several dismountable parts [21,22]. The latter are assembled before sintering and then disassembled when the component is sintered and ready to be ejected. A third approach is based on a tapered die solution that makes it possible to gradually decrease the radial pressure, thereby avoiding crack formation [23]. The tapered dies are difficult to manufacture for their complex geometrical shapes. A fourth approach consists of the die pre-stressing, representing a solution being well-known in metal forming. The die pre-stressing reduces the maximum tangential stress, known as the hoop stress, and the stress fluctuations, which the die is subject to. The use of the pre-stress for a sintering tool was justified by the possibility of reducing the internal diameter of the die during the process. Starting from the estimation of the radial pressure, the expected radial expansion of the sample in a free-die wall condition can be predicted [14]. The pre-stress was then designed to achieve an internal diameter reduction being equal to, or larger than, the sample expansion. After sintering, the stress-ring was taken off and the pre-stress was removed from the die which, thus, recovers to the original inner diameter. The radial pressure was decreased, thereby reducing the friction and ejection force. The die pre-stressing can be obtained using more than one approach e.g., stress-rings [24], stress-pins [25], stripwound containers [26], and a SMART® application using shape memory alloys (SMAs) [27,28]. Each of the die pre-stressing solutions features advantages and disadvantages [26,29–31].

This paper presents the development of a pre-stressed tool system for the manufacture of micro-gears [32]. The developed pre-stressed tool system worked according to HIP. Fourteen gears were produced using the pre-stressed tool system and then compared to other 14 gears produced without the stress-ring strategy. Different combinations of punch pressure, sintering temperature, and holding time were used to manufacture the gears. The comparison between the two tool systems was based on the ejection force and part fidelity to the computer-aided design (CAD) file. The ejection force was measured during the gear manufacture, while the part fidelity was documented using an optical microscope and computed tomography in order to obtain a multi-scale characterization. Furthermore, porosity analysis was carried out on the gears using computed tomography.

2. Materials and Methods

Figure 1 shows the aluminum micro-gear on which the tooling system was developed in this work. The sintered aluminum powder (AVL Metal Powders, Kortrijk, Belgium) was pure at 99.7%, with a fineness ranging from 42 to 250 μm. The gear had a circumscribed and inscribed circle of 3.5 mm and 2.55 mm, respectively. The height of the gear was of approximately 2 mm. The dimensional tolerances were set in the range of 0.005–0.010 mm. Since 2.7 g/cm^3 is the assumed aluminum density, the estimated powder weight is 0.038 ± 0.002 g.

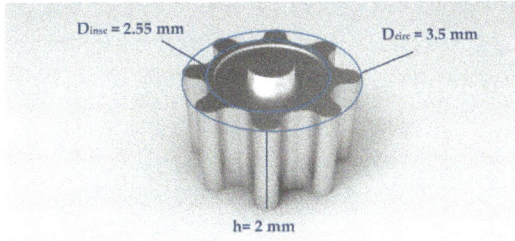

Figure 1. Experimental micro-gear design used in this work.

Figure 2 shows the tool system designed for producing the micro gears. The tool system included three elements: stress-ring, sleeve, and die section, see Figure 2a. The die was inserted into the sleeve, and then both were shrink-fitted with the stress-ring, see Figure 2b. The components of the tool were manufactured using electro discharge machining (EDM). The functional surfaces of the three elements of the tool system were all finished to bring the surface roughness down to Ra = 0.2–0.3 μm. The shape at the interference section between the sleeve and the stress-ring was conical to allow the assembling and releasing of the shrink-fit for several sintering cycles. Such a shape approach was based on the experimental investigations of the cold compaction of powder into an elastic die carried out by Noveanu et al. [33]. The tool material was H13 hot-work steel because of its mechanical properties at high temperatures.

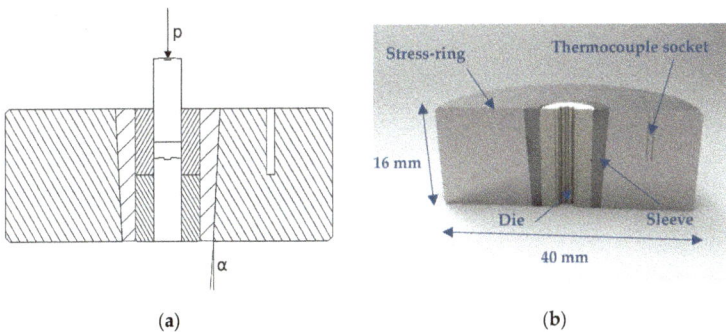

(a) **(b)**

Figure 2. Tool system design for the shrink-fit application: (a) stress-ring, sleeve and die section; and (b) rendered CAD model of the stress-ring, sleeve and die.

The main geometrical characteristics of the designed tool system were the average inner diameter of the stress-ring, $D_{stress\text{-}ring\ inn}$, the average outer diameter of the sleeve, $D_{sleeve\ out}$, the resulting interference, $I_\emptyset$, the pressure achieved at the sleeve/stress-ring interface, p_{int}, and the radial reduction obtained at the inner die diameter, D_{red}. The values of the geometrical characteristics were listed in Table 1.

Table 1. Dimensional specifications of the pre-stressed tool.

$D_{stress\text{-}ring\ inn}$	$D_{sleeve\ out}$	$I_\varnothing$	P_{int}	D_{red}
11.95 mm	12.00 mm	0.05 mm	326 MPa	12 µm

All the geometrical characteristics of the tool system were quantified according to the thick-walled hollow cylinder theory [34] and trade-off choices between the maximum yield strength of the tool material and the estimated total radial pressure arising from the sintering process. The total radial pressure was estimated taking into account the effect of the punch and sintering temperature during the sintering process. The gears were modelled as cylinders having an external diameter as the circumscribed circles. Such a simplification made it possible to design the tool system according to a worst-case pressure scenario. The first component of the radial pressure is the one generated by the punch compaction and was calculated according to Bockstiegel's model [15], resulting in a hysteresis cycle being shown in Figure 3. By assuming a maximum experimental pressure of 150 MPa for the axial compaction of the powder, the radial pressure at the end of the sintering process was quantified to be 128 MPa.

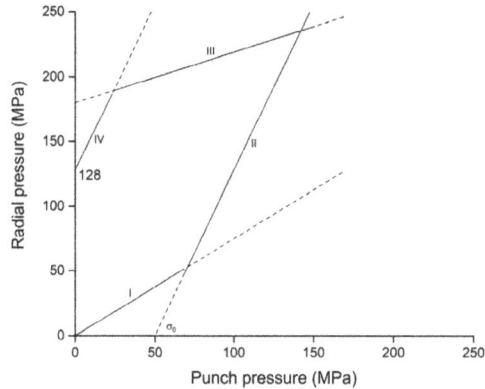

Figure 3. The diagram shows the estimated hysteresis cycle during the axial compaction of the investigated micro-gear according to the Bockstiegel model [15]. The diagram was obtained by assuming an aluminum-steel frictional condition µ = 0.61. Stages: I, elastic loading; II, plastic loading; III, elastic releasing; IV, plastic releasing.

The second component of radial pressure was originated by the difference in the thermal expansion between the gear and die. The thermal expansion equation was written using a differential way, dD, subtracting the two lengths involved for the die and the gear:

$$dD = \left| \left(\alpha_{gear} - \alpha_{die} \right) D_0 \Delta T \right| \tag{1}$$

Here α_{gear} and α_{die} are the thermal expansion coefficients of gear and die, respectively; D_0 is the nominal diameter at the die/gear interface; ΔT is the temperature variation with respect to the reference temperature of 20 °C. The maximum experimental sintering temperature was set to be 600 °C. The estimated dD corresponded to the theoretical expansion, which would have affected the gear in a wall free-die configuration. dD, therefore, represented a deformation in a closed-die. The estimated deformation of the gear generates a radial pressure of 260 MPa according to the Hooke's elasticity law. Since they were made of the same material, no difference in thermal expansion was assumed among the stress-ring, sleeve, and die.

The total radial pressure was quantified to be 388 MPa by summing the two above mentioned contributions, leading to a theoretical diameter expansion of the gear, D_{exp}, in a case of wall-free die [14] as follows:

$$D_{exp} = \frac{p_r - \nu p_r - \nu p_a}{E} D_0 \tag{2}$$

where p_r is the total radial pressure previously calculated; p_a is the axial compaction pressure; ν is the Poisson's ratio; E is the Young's modulus of aluminum. The D_{exp} value was found to be 13.48 μm. The nearest value to the D_{exp} value, complying with the yield limit of the used tool material, was chosen as the D_{red}.

Visual inspections and computed tomography inspections were carried out on the manufactured gears. Visual inspections were carried out to evaluate the impact of the reduction of the ejection force on the surface of the gears using an optical microscope DeMeet-220 (Schut Geometrical Metrology, Groningen, The Netherlands) with a 5:1 lens. The same lighting, fixturing process, and inspection area were used for all the inspected gears. The gears were gently cleaned before being inspected. Computed tomography (CT) inspections were, furthermore, carried out to evaluate the impact of the pre-stress on the resulting porosity and dimensional accuracy of the gears. CT is an imaging method that takes advantage of the capability of X-rays to penetrate the material [35,36].

3. Experimental Setup

Figure 4 shows the developed tool system and its assembly. The tool system was pre-stressed by assembling the die into the sleeve and mounting both into the stress-ring, as shown in Figure 4a. A mechanical press was used to assemble the tools and generate the required fit. The achieved diameter reduction was measured using the optical microscope and found to agree with the estimated value shown in Table 1 in a range of ±2 μm. After assembling, the whole tool was placed into a heating nozzle (Watlov, St. Louis, MO, USA) as shown in Figure 4b. The sintering process was carried out according to the HIP principle. The ejection force representing the parameter of interest in this work was measured using a load cell, U10M (HBM, Darmstadt, Germany).

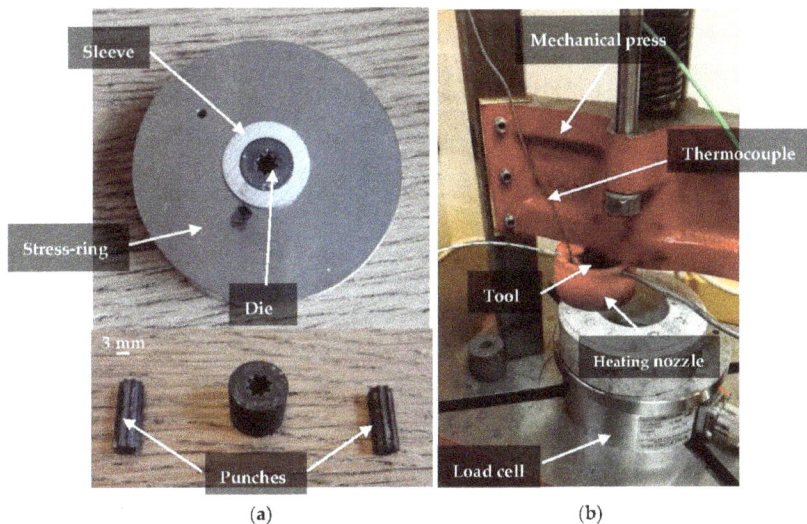

(a)   (b)

Figure 4. Experimental setup: (**a**) details of the overall pre-stressed tool system, die and punches; and (**b**) detail of the pre-stressed tool configuration during sintering.

The designed tool system was tested under different sets of process parameters such as punch pressure (four levels: 75 MPa, 100 MPa, 125 MPa, and 150 MPa), sintering temperature (four levels: 450 °C, 500 °C, 550 °C, and 600 °C), and holding time (four levels: 10 min, 15 min, 20 min, and 25 min). The highest punch pressure and sintering temperature represented the limit conditions to which the designed tool system should be subjected to. The tests were carried out using a partial factorial design, ensuring a randomization of the tests. The punch pressure and sintering temperature were monitored using a data acquisition system (DAQ), Figure 5a. A temperature control system allowed setting the desired sintering temperature for the process, see Figure 5b. The holding time was sampled by using a chronometer. The electronics of the setup was kept at an environment temperature of 21 ± 1 °C, ensuring its stability against thermal drifts. Moreover, the electronics of the setup was rebooted at a step of 25 min to avoid any additional drifts.

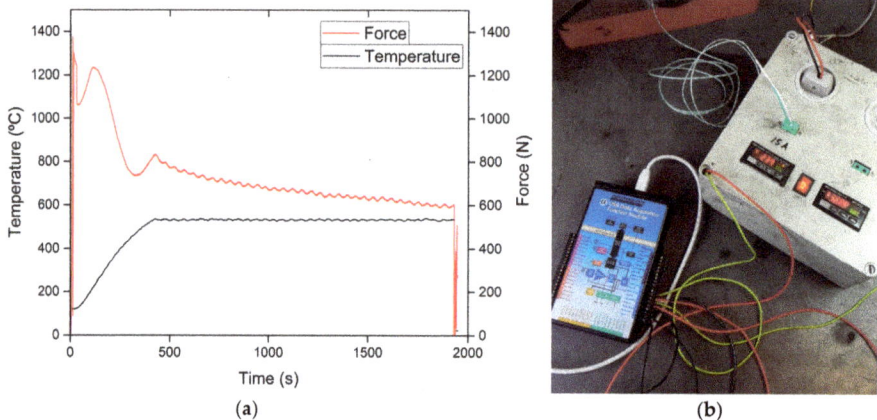

Figure 5. Measuring setup: (**a**) diagram showing the variation of the main process parameters; (**b**) data acquisition system, Data-Translation DT-9800 (Data Translation GmbH, Bietigheim-Bissingen, Germany), and temperature control system, Allen-Bradley 900 TC-32 (Rockwell Automation, Milwaukee, WI, USA).

Fourteen gears were produced using the pre-stressed tool system and coded as pre-stressed gears (PSGs) throughout the paper. Subsequently, the stress-ring was taken off and the other 14 gears were produced using the same procedures and process parameters used for PSGs. Such gears are coded as no pre-stressed gears (NPSGs) throughout the paper and used as reference conditions. By removing the stress-ring, a non-prestressed tool system with similar geometries and materials to the pre-stressed tool system was realized. The production of the two sets of gears was carried out in both cases using the same sets of sintering process parameters. Molybdenum disulfide (MoS_2) was used as a lubricant in order to avoid any problem arising from the possible aluminum powder adhering on the tools. Between the manufacture of two gears, the tool system was cooled down, ensuring that the same initial conditions were established for all tests. No effect of the tool wear was considered due to the limited amount of gears manufactured.

The uncertainty of the ejection force measurements was estimated according to the ISO Guide to the Expression of Uncertainty in Measurement (GUM) [37]. The following standard uncertainty contributions: load cell certificate, load cell repeatability when a reference standard was measured, press repeatability, load cell resolution, and eccentric load were considered and modelled using a rectangular distribution [37]. Neither linearity, nor hysteresis, was taken into account because the used measuring range was smaller than 1% of the total measuring range of the used load cell (500 kN). The environment temperature fluctuations were measured, found to be small, compared to other contributions and, thus, neglected. The expanded measurement uncertainty value, with a coverage factor k = 2 at 95% confidence level, was estimated to be in the range of 10–100 N. The largest source of uncertainty was found to be the press repeatability, which was strongly dependent on the process parameters.

4. Results and Discussions

Figure 6 shows an example of gears produced using the developed tool system. Figure 7a–c show the evolution of the measured ejection force under different testing conditions. The three series of experimental results were obtained with the punch pressure, sintering temperature and holding time varying once at a time. Each force value was expressed in terms of the average value and expanded measurement uncertainty. By comparing the ejection forces measured between NPSGs and PSGs, it was seen that a strong reduction of the ejection force was achieved for all the considered sintering sensations. The reduction of the ejection force was found to be several times larger than the expanded uncertainty of the ejection force measurements, giving evidence that the reduction had a physical fundament and, thus, can be replicated over the time.

Figure 6. Examples of sintered samples using different process parameters. A pen tip was used to give an idea of the real size of the micro-gears.

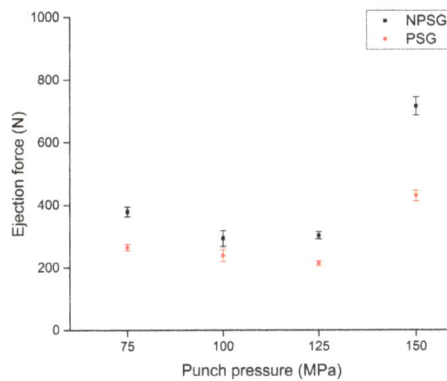

(a)

Figure 7. *Cont.*

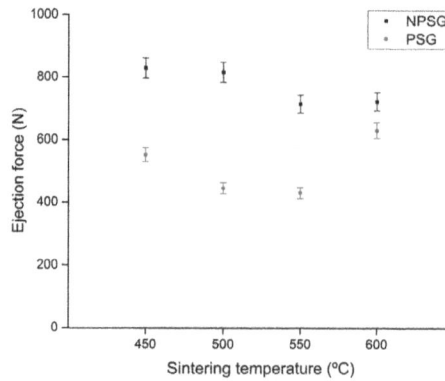

(b)

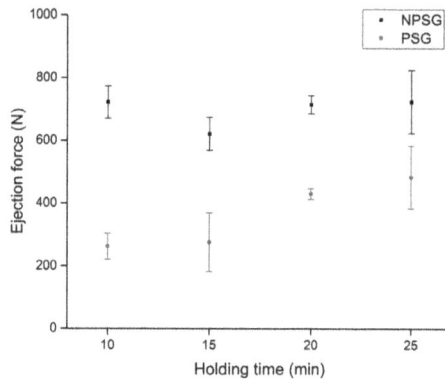

(c)

Figure 7. Diagrams showing the experimental values of ejection force measured during the ejection of the micro-gears with respect to the process parameters. (**a**) The four level of punch pressure with constant sintering temperature (550 °C) and holding time (20 min); (**b**) the four level of sintering temperatures with constant punch pressure (150 MPa) and holding time (20 min); and (**c**) the four level of holding time with constant punch pressure (150 MPa) and sintering temperature (550 °C).

In the case of the PSGs, the lower ejection forces were achieved because of the radial pressure reduction obtained after removing the stress-ring from the tooling system. As a consequence, the die expanded to its designed dimensions, thereby generating a clearance between the tool and gear. Such a clearance resulted in the reduction of the friction and ejection force. The average percentage reduction, F_{red}, was calculated for each experimental series, i.e., punch pressure, sintering temperature, and holding time, as follows:

$$F_{red}[\%] = \frac{\sum_{i=1}^{n}\left(\dfrac{F_{NPSG\,(i)} - F_{PSG\,(i)}}{F_{NPSG\,(i)}} \cdot 100\right)}{n} \tag{3}$$

where n is the number of comparisons made for the considered experimental series; F_{NPSG} and F_{PSG} are the ejection forces measured for the micro-gears sintered with the same process parameters by using a configuration without and with pre-stress, respectively. A summary of the average results for each experimental series was collected in Table 2. The holding time series showed a reduction of the advantage of the pre-stress as the holding time increased. The longer holding times were, the larger the ejection forces. Such a finding can be due to the fact that longer holding times increased the adhesion of a gear to the die walls, resulting in stronger interactions between the surfaces in contact. A similar percentage force reduction was observed by varying either the punch pressure or the sintering temperature.

Table 2. Average ejection force reduction during the ejection as a function of the process parameters.

Punch Pressure	Sintering Temperature	Holding Time
−27%	−33%	−50%

4.1. Visual Inspection of Micro-Gears Using an Optical Microscope

Figure 8 shows a collection of pictures of the gears sintered in the present work. PSGs did not show any ploughing effect compared to NPSGs where such effect was well visible, as shown in the figure using arrows. Ploughing was caused by asperities of a tool material metal penetrating into a gear material, which was the softer one. Although it may be deducted that the new developed tool system increased the uniformity of the surface of the gears, the visual inspection did not give sufficient evidences to make a definitive statement. As a consequence, a quantitative analysis was carried out using computed tomography.

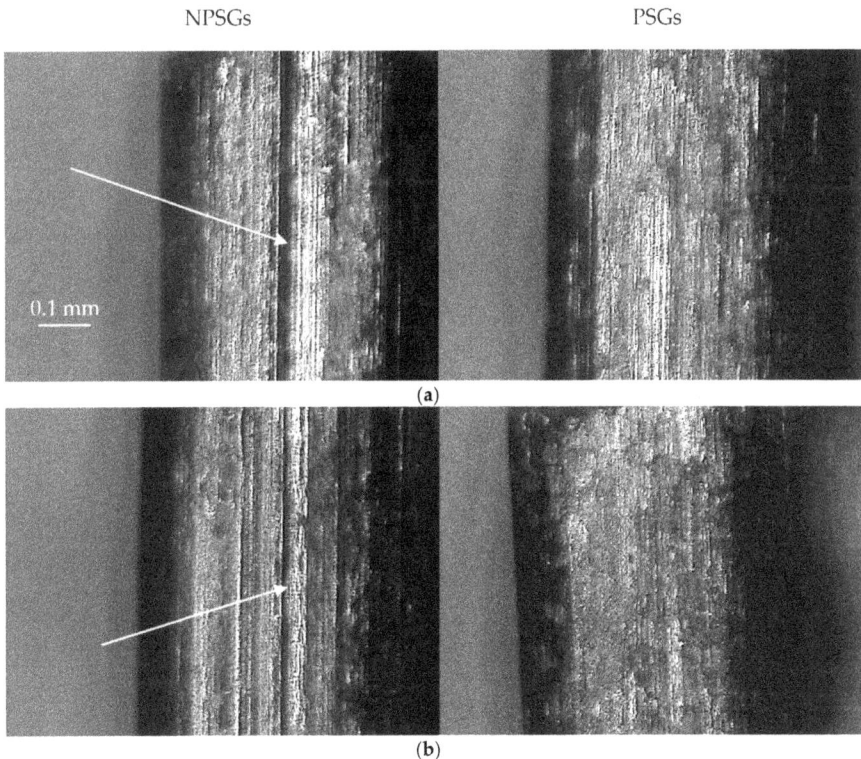

NPSGs   PSGs

(a)

(b)

Figure 8. *Cont.*

9

(c)

Figure 8. The pictures showing the lateral surfaces of NPSGs, left side, and PSGs, right side, acquired using an optical microscope. PSGs: (**a**) T: 550 °C, holding time: 10 min, punch pressure: 150 MPa, percentage force reduction: −66%; (**b**) T: 550 °C, holding time: 15 min, punch pressure: 150 MPa, percentage force reduction: −64%; and (**c**) T: 550 °C, holding time: 15 min, punch pressure: 150 MPa, percentage force reduction: −30%. Regardless of the sintering process parameters, NPSGs show on their surfaces the consequences of the larger ejections.

4.2. Characterization of Micro-Gears Using Computed Tomography

Five repeated scans of a PSG, and five repeated scans of a NPSG were carried out using a Nikon XT H 225 (Nikon, Tokyo, Japan) industrial CT. The two gears were manufactured subsequently in order to avoid any misjudgment due to wear. Table 3 shows an overview of the scanning parameters used in this work. A power of 6 W and a voxel size of 2.8 μm were used for the scanning process, leading to an isotropic scanning resolution of approximately 4 μm. The selected resolution required to minimize the rotary table wobble using a large number of projections and reversal scanner movements designed by the authors. The detector calibration was performed by using a total of 256 projections evenly distributed over four different power levels, with 1024 projections per level. Four power levels were sufficient due to the constant absorption values of the aluminum across the entire spectrum. Due to the high resolution selected, it was necessary to stabilize the used CT for 2 h, otherwise thermal drifts would have impaired the CT measurements. The reconstruction of the projections into a 3D volume was finalized using Nikon Metrology CT PRO 3D version 3.1.9 (Nikon, Tokyo, Japan) without applying any software-based correction features to cope with scanning errors. Surface determination was based on a local thresholding method, implemented in VG studio Max 3.0 (Volume Graphics, Heidelberg, Germany). This method currently represents the state-of-the-art tool for segmenting CT datasets, providing an accuracy of up to 1/10 of the voxel size [38]. The surface was generated by manually selecting the grey values belonging to the gear and to the background on the reconstructed images as seen from different viewpoint angles. A grey value represents the intensity of a voxel [35] and is expressed in terms of shades of grey [39]. Three off-line reference artifacts for systematic errors correction were used to compress the systematic errors and to establish traceability in this work. An off-line reference artefact with a calibrated sphere of 150 μm was used to establish traceability of the porosity and dimensional analysis. Two off-line reference artifacts based on several spheres of 2 mm were used to establish traceability of the dimensional analysis. Since high quality X-ray projections are an essential prerequisite for accurate CT measurements, an image quality analysis was carried out on the X-ray projections. The extent of image noise was quantified and found to be smaller than 0.4% of the average voxel intensity. Furthermore, no missing frequencies were observed in the reconstructed volumes of the gears, guaranteeing that no Feldkamp artifacts [39] compromised the CT measurements. The frequency analysis was performed in the Fourier domain using Fiji post-processing software.

Table 3. Overview of the scanning parameters used in this work.

Parameter	Unit	Value
X-ray tube voltage	kV	130
X-ray tube current	μA	50
Corrected voxel size	μm	2.8
Magnification factor	-	60
No. of projections	-	2000
No. of image per projection	-	2
Integration time	s	0.5
Scanning time	min	34

The porosity analysis was performed using two representative 3D volumes, one volume per gear, to minimize the impact of external influences such as dust, image noise, and lubricant. The representative 3D volumes were set to enclose approximately 90% of the total volume of each gear. It was, therefore, assumed that any result obtained from the inspection of the two volumes was representative of the two entire gears. By using the Procedure for Uncertainty MAnagement (PUMA) method [37], the expanded measurement uncertainty, at 95% confidence level, for the porosity analysis was estimated to be equal to ± 0.01 mm^3. The expanded measurement uncertainty was based on the following uncertainty contributions: (i) the uncertainty from the calibration certificates of the off-line reference artefact; (ii) the uncertainty from the measurement repeatability; and (iii) the uncertainty arising from the image noise. All the uncertainty contributions were based on Type B evaluations [37]. The image noise contributed 60% of the expanded measurement uncertainty. Figure 9 shows the spatial porosity distribution for the two gears inspected using CT. The porosity fraction was quantified to be 0.03 ± 0.01% and 0.05 ± 0.01% of the representative volume for PSG and NPSG, respectively. By performing a *t*-test [40], it was found that the two porosity distributions can be assumed to be statistically different, leading to a conclusion that the pre-stress increased the compactness of the powder of which the gears were composed. The *t*-test was conducted at the 95% confidence level and took into account the measurement uncertainty of the porosity fractions. It was also observed that the pores were found to be distributed within the volumes of both selected gears in a different fashion. PSG showed a uniform porosity within the inspected volume, whereas NPSG did not. As a general conclusion, the use of pre-stress reduced the gear porosity by 40% and allowed the achievement of a more uniform spatial distribution of the pores in the gear.

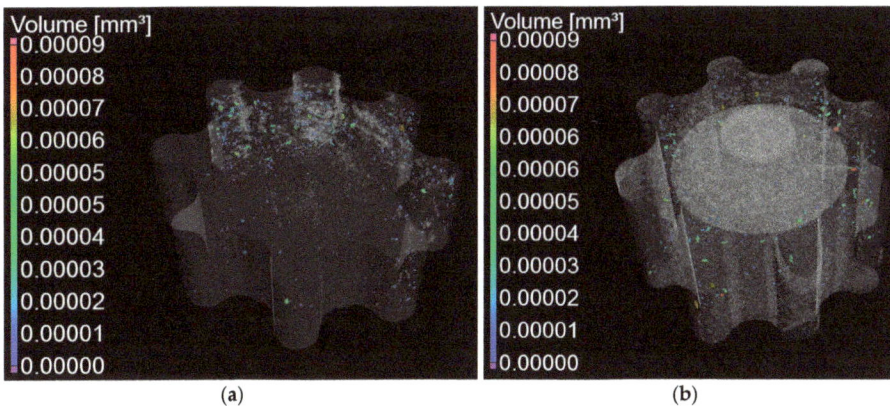

Figure 9. Volume pore distributions for (**a**) NPSG and (**b**) PSG. Deviation range from 0 to 0.00009 mm^3.

The dimensional analysis of the gears was based on the actual-to-nominal comparison tool [40]. The use of the actual-to-nominal comparison allows the quantification of the deviation of each surface point of an object compared to its CAD file [36]. An actual-to-nominal comparison simplifies the measurement procedures compared to other approaches involving primitive features [36]. By using the PUMA method [37], the expanded measurement uncertainty per voxel was calculated, making each point of an actual-to-nominal comparison traceable to the meter. The measurement uncertainty was based on the following uncertainty contributions: (i) the uncertainty from the calibration certificates of the two used off-line reference artifacts; (ii) the uncertainty from the measurement repeatability; (iii) the uncertainty arising from the image noise; and (iv) uncertainty from the post-processing activities. The expanded measurement uncertainty was found to be 0.003 mm for all the voxels of the reconstructed volumes. A single uncertainty value was sufficient in this case due to the following favorable conditions: (i) no gear movements occurred during scanning, establishing that no local reconstruction errors occurred; (ii) the absence of material-related image artifacts, ensuring that X-rays were well modelled during reconstruction at any voxel; (iii) the axial symmetry of the gears, making sure that scatter and the X-ray absorption coefficient did not change locally. The measurement uncertainty per voxel represents a source of novelty, paving the way for 3D measurement uncertainty estimations. Figure 10 shows the two actual-to-nominal comparisons for the two gears with the deviation range and colors. It can be seen from the figures that the two gears appeared to be extremely different, with PSG having the higher fidelity to the CAD file. PSG showed 95% of its surface lied within ±0.009 mm, whereas PSG showed that 95% of its surface lies within ±0.040 mm. By using a *t*-test [40], it was confirmed that the two fidelity values were statistically different. As a general conclusion, the adoption of a pre-stress value can improve the dimensional accuracy of gear by more than 75% compared to gears produced without pre-stress.

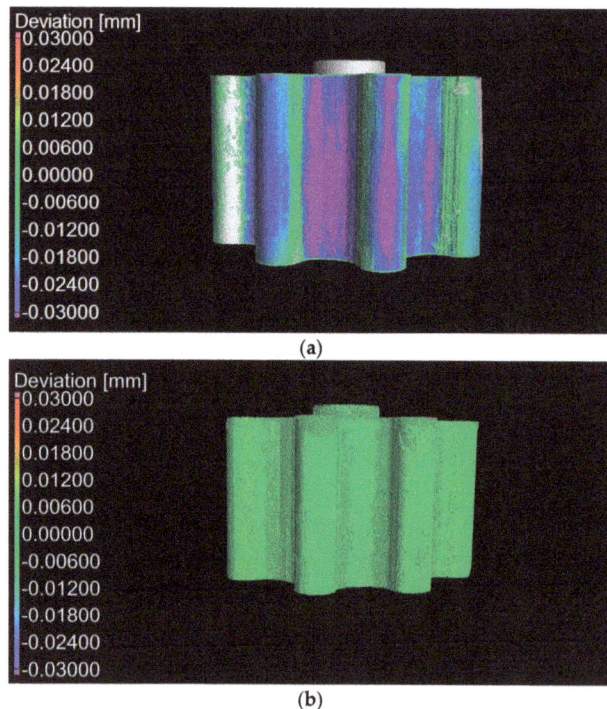

Figure 10. Actual-to-nominal comparison for (**a**) NPSG and (**b**) PSG. Deviation range ±0.03 mm.

5. Conclusions

This paper has presented the development of a pre-stressed tool system for the manufacture of micro gears made of aluminum [32]. The developed pre-stressed tool system included a stress-ring and worked according to HIP, thereby reducing the sintering time compared to conventional sintering. Fourteen gears were produced using the pre-stressed tool system and then compared to another 14 gears produced after removing the stress-ring. All the gears were manufactured in an environment kept at of 21 ± 1 °C and using different combinations of punch pressure, sintering temperature, and holding time. By using the newly-developed pre-stressed tool system, a reduction in the ejection force of up to 50% was observed. The magnitude of the ejection force reduction was found to be larger than the ejection force measurement uncertainties, giving evidence that such results had a physical explanation and, thus, can be replicated over time. Visual inspections, carried out using an optical microscope, showed a reduced frictional interaction on the lateral surface of the micro-gears produced by using a pre-stressed tool system. Furthermore, CT measurements documented that the new tool system reduced the porosity in the gear by 40% and improved the dimensional fidelity to CAD by more than 75% compared to gears produced without pre-stress. Finally, the paper has also introduced the concept of CT measurement uncertainty per voxel, as well as proved the applicability of CT to micro-manufacturing with expanded uncertainties of a few microns.

Acknowledgments: This research work was undertaken in the context of MICROMAN project ("Process Fingerprint for Zerodefect Net-shape MICROMANufacturing", http://www.microman.mek.dtu.dk/). MICROMAN is a European Training Network supported by Horizon 2020, the EU Framework Programme for Research and Innovation (Project ID: 674801). The first author would also like to thank IPU and Micro-FAST [41] for the possibility to carry out experiments and for funding tools and dissemination activities related to this project. Moreover, special thanks go to Mogens Arentoft for his supervision and suggestions during the initial period of this research. Finally, special thanks go to the Imaging Industry Centre at DTU for making the CT scanner available.

Author Contributions: Emanuele Cannella designed the tools, operated the experiments, analyzed the results and wrote the paper. Emil Krabbe Nielsen designed the micro-gear shape, supported the experiments, and reviewed and edited the paper. Alessandro Stolfi performed all the CT measurements shown in this work, and reviewed and edited the paper.

Conflicts of Interest: The authors declare no conflict of interest.

References

1. Castro, R.H.R. Overview of Conventional Sintering. In *Sintering*; Castro, R.H.R., van Benthem, K., Eds.; Springer: Berlin, Germany, 2013; Volume 35, pp. 1–16.
2. Groza, J.R.; Zavaliangos, A. Nanostructured bulk solids by field activated sintering. *Rev. Adv. Mater. Sci.* **2003**, *5*, 24–33.
3. Wang, G.P.; Liu, W.Q.; Huang, Y.L.; Ma, S.C.; Zhong, Z.C. Effects of sintering temperature on the mechanical properties of sintered NdFeB permanent magnets prepared by spark plasma sintering. *J. Magn. Magn. Mater.* **2014**, *349*, 1–4. [CrossRef]
4. Garay, J.E. Current-Activated, Pressure-Assisted Densification of Materials. *Annu. Rev. Mater. Res.* **2010**, *40*, 445–468. [CrossRef]
5. Guillon, O.; Gonzalez-Julian, J.; Dargatz, B.; Kessel, T.; Schierning, G.; Rathel, J.; Herrmann, M. Field-assisted sintering technology/spark plasma sintering: Mechanisms, materials, and technology developments. *Adv. Eng. Mater.* **2014**, *16*, 830–849. [CrossRef]
6. Li, W.; Olevsky, E.A.; McKittrick, J.; Maximenko, A.L.; German, R.M. Densification mechanisms of spark plasma sintering: Multi-step pressure dilatometry. *J. Mater. Sci.* **2012**, *47*, 7036–7046. [CrossRef]
7. Akarachkin, S.A.; Ivashutenko, A.S.; Martyushev, N.V. Activation of mass transfer processes at spark plasma sintering of zirconium dioxide. *IOP Conf. Ser. Mater. Sci. Eng.* **2016**, *124*, 12042. [CrossRef]
8. Zhao, J.; Qin, Y.; Huang, K.; Hijji, H. Forming of micro-components by electrical-field activated sintering. *MATEC Web Conf.* **2015**, *21*. [CrossRef]
9. Orrù, R.; Licheri, R.; Locci, A.M.; Cincotti, A.; Cao, G. Consolidation/synthesis of materials by electric current activated/assisted sintering. *Mater. Sci. Eng. R Rep.* **2009**, *63*, 127–287. [CrossRef]

10. Munir, Z.A.; Quach, D.V.; Ohyanagi, M. Electric current activation of sintering: A review of the pulsed electric current sintering process. *J. Am. Ceram. Soc.* **2011**, *94*, 1–19. [CrossRef]
11. Delaizir, G.; Bernard-Granger, G.; Monnier, J.; Grodzki, R.; Kim-Hak, O.; Szkutnik, P.D.; Soulier, M.; Saunier, S.; Goeuriot, D.; Rouleau, O.; et al. A comparative study of Spark Plasma Sintering (SPS), Hot Isostatic Pressing (HIP) and microwaves sintering techniques on p-type Bi_2Te_3 thermoelectric properties. *Mater. Res. Bull.* **2012**, *47*, 1954–1960. [CrossRef]
12. Giuntini, D.; Olevsky, E.A.; Garcia-Cardona, C.; Maximenko, A.L.; Yurlova, M.S.; Haines, C.D.; Martin, D.G.; Kapoor, D. Localized overheating phenomena and optimization of spark-plasma sintering tooling design. *Materials* **2013**, *6*, 2612–2632. [CrossRef]
13. Chawake, N.; Pinto, L.D.; Srivastav, A.K.; Akkiraju, K.; Murty, B.S.; Kottada, R.S. On Joule heating during spark plasma sintering of metal powders. *Scr. Mater.* **2014**, *93*, 52–55. [CrossRef]
14. Long, W.M. Radial pressures in powder compaction. *Powder Metall.* **1960**, *6*, 73–86. [CrossRef]
15. Bockstiegel, G. The Porosity-Pressure Curve and its Relation to the Size Distribution of Pores in Iron Powder Compacts. In Proceedings of the 1965 International Powder Metallurgy Conference, New York, NY, USA, 14–17 June 1965.
16. Höganäs AB. Production of Sintered Components. Available online: https://www.hoganas.com/globalassets/media/sharepoint-documents/HandbooksAllDocuments/Handbook2_Production_of_Sintered_Components_December_2013_0675HOG_interactive.pdf (accessed on 5 June 2017).
17. Rahman, M.M.; Nor, S.S.M. An experimental investigation of metal powder compaction at elevated temperature. *Mech. Mater.* **2009**, *41*, 553–560. [CrossRef]
18. Enneti, R.K.; Lusin, A.; Kumar, S.; German, R.M.; Atre, S.V. Effects of lubricant on green strength, compressibility and ejection of parts in die compaction process. *Powder Technol.* **2013**, *233*, 22–29. [CrossRef]
19. Li, Y.Y.; Ngai, T.L.; Zhang, D.T.; Long, Y.; Xia, W. Effect of die wall lubrication on warm compaction powder metallurgy. *J. Mater. Process. Technol.* **2002**, *129*, 354–358. [CrossRef]
20. Luo, S.D.; Yang, Y.F.; Schaffer, G.B.; Qian, M. Warm die compaction and sintering of titanium and titanium alloy powders. *J. Mater. Process. Technol.* **2014**, *214*, 660–666. [CrossRef]
21. Chen, P.; Kim, G.-Y.; Ni, J. Investigations in the compaction and sintering of large ceramic parts. *J. Mater. Process. Technol.* **2007**, *190*, 243–250. [CrossRef]
22. Andresen, H.; Lund, E. Tooling Solutions for Cold and Warm Forging Applications for Automotive and Other Segments. Available online: http://www.istma.org/istma-world/ISTMA_Conferencehall/uddeholm2008/Tooling%20solutions%20for%20cold%20and%20warm%20forging_Erik%20Lund_Henrik%20Andresen.pdf (accessed on 5 June 2017).
23. Garner, S.; Ruiz, E.; Strong, J.; Zavaliangos, A. Mechanisms of crack formation in die compacted powders during unloading and ejection: An experimental and modeling comparison between standard straight and tapered dies. *Powder Technol.* **2014**, *264*, 114–127. [CrossRef]
24. Armentani, E.; Bocchini, G.F.; Cricrì, G. Doubly shrink fitted dies: Optimisation by analytical and FEM calculations. *Powder Metall.* **2012**, *55*, 130–141. [CrossRef]
25. Koç, M.; Arslan, M.A. Design and finite element analysis of innovative tooling elements (stress pins) to prolong die life and improve dimensional tolerances in precision forming processes. *J. Mater. Process. Technol.* **2003**, *142*, 773–785. [CrossRef]
26. Groenbaek, J.; Nielsen, E.R. Stripwound containers for combined radial and axial prestressing. *J. Mater. Process. Technol.* **1997**, *71*, 30–35. [CrossRef]
27. Pan, W.; Qin, Y.; Law, F.; Ma, Y.; Brockett, A.; Juster, N. Feasibility study and tool design of using shape memory alloy as tool-structural elements for forming-error compensation in microforming. *Int. J. Adv. Manuf. Technol.* **2008**, *38*, 393–401. [CrossRef]
28. Qin, Y. Forming-tool design innovation and intelligent tool-structure/system concepts. *Int. J. Mach. Tools Manuf.* **2006**, *46*, 1253–1260. [CrossRef]
29. Fu, M.W.; Chan, W.L. *Micro-Scaled Products Development via Microforming*; Fu, M.W., Chan, W.L., Eds.; Springer: Berlin, Germany, 2014; ISBN 9781447163268.
30. Ghassemali, E.; Tan, M.J.; Jarfors, A.E.W.; Lim, S.C.V. Progressive microforming process: Towards the mass production of micro-parts using sheet metal. *Int. J. Adv. Manuf. Technol.* **2013**, *66*, 611–621. [CrossRef]
31. Johnson, D.; Martynov, V.; Gupta, V. Applications of shape memory alloys: Advantages, disadvantages, and limitations. *Proc. SPIE* **2001**, *4557*, 341–351. [CrossRef]

32. Cannella, E.; Nielsen, E.K.; Arentoft, M. Ejection force analysis of sintered aluminium micro gears using a shrink-fit die principle. In Proceedings of the 11th International Conference on Multi-Material Micro Manufacture (4M2016): Co-organised with 10th International Workshop on Microfactories (IWMF2016), Kgs. Lyngby, Denmark, 13–15 September 2016; pp. 41–44.

33. Noveanu, D. Researches concerning a new method for obtaining spur gears by metal powder compaction in elastic dies. *Metalurgia* **2013**, *65*, 35–39.

34. Atzori, B. Gusci spessi. In *Appunti di Costruzione di Macchine*; Libreria Cortina: Padova, Italy, 2001; pp. 262–281.

35. Kruth, J.P.; Bartscher, M.; Carmignato, S.; Schmitt, R.; De Chiffre, L.; Weckenmann, A. Computed tomography for dimensional metrology. *CIRP Ann. Manuf. Technol.* **2011**, *60*, 821–842. [CrossRef]

36. De Chiffre, L.; Carmignato, S.; Kruth, J.P.; Schmitt, R.; Weckenmann, A. Industrial applications of computed tomography. *CIRP Ann. Manuf. Technol.* **2014**, *63*, 655–677. [CrossRef]

37. *Geometrical Product Specifications (GPS)—Inspection by Measurement of Workpieces and Measuring Equipment—Part 2: Guidance for the Estimation of Uncertainty in GPS Measurement, in Calibration of Measuring Equipment and in Product*; ISO 14253-2:2011; British Standards Institute: London, UK, 2011.

38. Müller, P. *Coordinate Metrology by Traceable Computed Tomography*; Technical University of Denmark (DTU): Kgs. Lyngby, Denmark, 2013.

39. Stolfi, A. Integrated Quality Control of Precision Assemblies Using Computed Tomography. Ph.D. Thesis, Technical University of Denmark (DTU), Kgs. Lyngby, Denmark, 2017.

40. Witt, P.L.; McGrain, P. Comparing two sample means t tests. *Phys. Ther.* **1985**, *65*, 1730–1733. [CrossRef] [PubMed]

41. MICRO-FAST. Available online: www.micro-fast.eu (accessed on 5 June 2017).

micromachines

MDPI

Article

Rapid Fabrication of Disposable Micromixing Arrays Using Xurography and Laser Ablation

J. Israel Martínez-López *, H.A. Betancourt, Erika García-López, Ciro A. Rodriguez and Hector R. Siller

Tecnológico de Monterrey, Eugenio Garza Sada 2501 Sur, 64849 Monterrey, NL, Mexico; habc-8@hotmail.com (H.A.B.); garcia.erika@itesm.mx (E.G.-L.); ciro.rodriguez@itesm.mx (C.A.R.); hector.siller@itesm.mx (H.R.S.)
* Correspondence: israel.mtz@itesm.mx; Tel.: +52-818-358-2000

Academic Editors: Guido Tosello, Hans Nørgaard Hansen and Nam-Trung Nguyen
Received: 28 February 2017; Accepted: 28 April 2017; Published: 4 May 2017

Abstract: We assessed xurography and laser ablation for the manufacture of passive micromixers arrays to explore the scalability of unconventional manufacture technologies that could be implemented under the restrictions of the Point of Care for developing countries. In this work, we present a novel split-and-recombine (SAR) array design adapted for interfacing standardized dispensing (handheld micropipette) and sampling (microplate reader) equipment. The design was patterned and sealed from A4 sized vinyl sheets (polyvinyl chloride), employing low-cost disposable materials. Manufacture was evaluated measuring the dimensional error with stereoscopic and confocal microscopy. The micromixing efficiency was estimated using a machine vision system for passive driven infusion provided by micropippetting samples of dye and water. It was possible to employ rapid fabrication based on xurography to develop a four channel asymmetric split-and-recombine (ASAR) micromixer with mixing efficiencies ranging from 43% to 65%.

Keywords: micromixing; split and recombine; rapid manufacture; xurography; laser ablation

1. Introduction

The development and spread of advanced diagnostic devices based on microfluidic technology can be a key strategy element to tackle a wide variety of infectious diseases such as AIDS, tuberculosis (TB) or malaria. Despite the availability of diagnostics and solutions for these diseases, every year about 15 million people die from these diseases [1]. Academics have proposed a wide variety of devices built on advanced manufacturing, electromechanical sensors and actuators, and Information and Communications Technologies (ICT) to treat patients outside the boundaries of a hospital. A significant portion of these devices, gadgets, and systems had been conceived to serve in the Point of Care (POC). The approach to bringing healthcare closer to the patients is inherently well encompassed with technologies that enable the miniaturization of established diagnostics, treatment, and monitoring illness. In developing countries, mobility is an asset that can enhance aid by providing closer access to remote areas, treating illnesses at earlier stages, and deterring the spread of infectious diseases.

The World Health Organization (WHO) has resumed the requirements of the in-field solutions under the acronym "ASSURED", which stands for affordable, sensitive, specific, user-friendly, rapid and robust, equipment-free or minimal equipment, and deliverable to end-users. The design and manufacture of POC devices carries added demanding development requirements. This equipment should remain functional in its main purpose reliably under the constraints of being low-cost, the absence of trained staff, lack of electricity, poorly equipped laboratory facilities, and limited access to refrigeration and storage [2,3]. Studies evaluating the performance of POC prognosis have shown that end-users (typically remote health workers or volunteers) are affected by some factors, including manual dexterity, visual acuity, and available lighting during testing. Micropipette or other standardized methods for sample management and transfer can support a more reliable and high-throughput capable device operation [4,5].

Despite the copious amount of research in the design and operation of micromixers, research regarding manufacture technology towards implementation beyond academic environments is limited. Microfabrication-based on the photolithographic processes of polymers is the most common approach employed [6]. Typically, this implies procedures that require supervision by specialized personnel under laboratory facilities (i.e., spin coaters, ovens, plasma treatment, hot plates) and requires the supply of resins for the development of structures within the finesse in the micrometric scale. These specimen can then be used as sample devices or employed to replicate them through casting, stamping, or injection molding. Polydimethylsiloxane (PDMS) is the most popular elastomer due the ability to cast with nanometric resolution and to be relatively inexpensive and because it can be irreversibly bonded to other materials such as glass or other polymeric films [7,8].

In rapid fabrication manufacturing techniques, devices are fabricated faster than conventional manufacturing processes. While some of these methodologies were originally conceived solely for producing one or few samples within some surface or structural quality limitations, now it is possible to find cases where these techniques have evolved into the production of components that could not be made otherwise [9]. The incorporation of rapid fabrication technology for microfluidic devices is a growing trend among researchers, but it has not yet been fully developed [10]. 3D printing has raised much awareness among academics and media because it allows devices to be manufactured on-demand with ease and quickness for medical applications [11] of microfluidic devices [12].

Compared to other rapid fabrication technologies such as fused deposition modeling (FDM) or stereolithography (SL), technology that relies on processing thin-films rolls is typically more affordable and can be transported more easily than liquid-based reagents like PDMS or the resins used for the processes mentioned above. The roll to roll hot embossing process is a recent advancement in micro hot embossing processes and is capable of continuously fabricating micro/nano structures on the polymer with high efficiency and high throughput [13,14]. High volume fabrication employing manufacture based on processing polymer on rolls has been used successfully for pinched flow fractionation on cellulose acetate [13]. Another group developed an electrophoresis chip for the detection of antibiotic resistance bacteria by feeding a thermoplastic foil through a hot embossing cylinder [15]. Senkbeil et al. have altered the rheological behavior of the resin system to produce capillary electrophoresis chips.

Xurography is a rapid manufacture technology that originated from the adaptation of equipment intended for advertisement used for the development of microfluidic systems. It relies on the patterning and removal of thin-films materials using a blade tracing a design [16,17]. Originally, around a decade ago, the major advantage was the simplicity and quickness in operation; nowadays, the cost reduction in these equipment from the original 4000 USD to 200 USD has accentuated the price-value ratio advantage of this technique compared to more conventional approaches. For example, Silhouette [18] and Cricut [19] are two providers that offer desktop sized cutting equipment designed for home use with a starting price of 200 USD for processing standardized A4 sheets with comparable cutting performance to more expensive equipment.

Another technique used for manufacturing microdevices from thin layers of materials formatted as rolls is laser ablation. Some research has been developed widely in PDMS [20], glass [21,22], cyclic olefin copolymer (COP) [23], and polymethyl methacrylate (PMMA) [24,25] materials with a variety of applications in lab on a chip field. These microdevices have been processed using an ultra short pulsed laser (e.g., femtosecond and picosecond pulse durations), which results in a promising technique in micromachining that relies on design flexibility, precision, and productivity. Laser ablation for microdevices is performed through the interaction between laser energy and the material, where the main parameters are wavelength and pulse duration. Therefore a focused volume absorbs laser energy, which allows a localized machining, while the rest of the sample results are unaffected [26]. Among the laser ablation equipment, engravers are a subset of machines suited for cutting non-metallic materials.

Prices can vary widely depending on the type of laser, work area, and features from roughly 8000 USD to 5000 USD. Recently there have been some efforts funded by crowdfunded projects to commercialize prototypes focused on hobbyists with lower price baselines and added features [27,28]. However these alternatives remain above 500 USD and are still in development.

One of the desirable operations that a POC device must perform is micromixing since devices at the micrometric scale tend naturally to operate on a laminar flow regime. Under that condition, the homogenization and reaction of reagents tend to be slow or require systems with large characteristic length to function properly [29,30]. Rather than perform micromixing by a supplementary force, passive micromixing depends on geometrical features along the flow chamber. For example, slanted grooved [31,32] and staggered herringbone [33] micromixers induce homogenization by creating secondary flows using obstacles or other complex features along the flow chamber. These designs are typically highly efficient in the task of mixing but also require expensive multi-layer manufacture technology.

A widely studied example of a passive micromixer is the T-mixer. In this simple design, two separate fluids are brought into contact from opposite directions and then leave through a channel that is perpendicular to the inlet channels [24]. The performance tends to be low as the mixing occurs only proximate to the junction. A more recent methodology, SAR or ASAR micromixers force this contact by repeatedly putting together the streams from the inputs and hence increasing the interfacial area of the streams. Recent research had employed the principle of the T-mixer by splitting and recombining streams iteratively with more complex geometry as rhombic [34,35], modified Tesla [36–38], or curve based [39–41] and shapes based setups. Table 1 represents the recent development and the manufacture methodologies and some micromixers channels per device (N) for in-plane micromixers prior to this work. It is noticeable that, before the present work, Chung et al. [42] used a laser as part of a manufacturing process, but this was not done with consideration of either POC or rapid manufacturing. The authors have not found previous work from other research groups on the development of interfacing a micromixer with standardized sample management equipment, neither have they found deported efforts to develop full-sized arrays.

Table 1. Manufacture methodologies for in-plane micromixers.

Work	Reference	Manufacture Methodology	N	w_{input}
Hong et al. (2004)	[37]	Molding (nickel-SU-8), photolitography, hot embossing, drilling, thermal bonding	1	200 μm
Sudarsan & Ugaz (2006)	[39]	Circuit printing, etching, heat treatment	1	150 μm
Chung & Shi (2007)	[34]	Lithography, micro-molding, oxygen plasma treatment bonding, mechanical punching	1	500 μm
Chung et al. (2009)	[42]	Laser machining, PDMS casting from PMMA, thermal and oxygen plasma bonding, mechanical punching	1	500 μm
Ansari et al. (2010)	[40]	SU-8 photolithography over a silicon wafer, PDMS molding, mechanical punching	1	300 μm
Scherr et al. (2012)	[43]	SU-8 photolithography, PDMS molding, plasma cleaning, mechanical punching	1	30–200 μm
Li et al. (2013)	[44]	PDMS molding	1	300 μm
Martínez-López et al. (2016)	[10]	Xurography of PVC and manual lamination	1	750 μm

To address these restrictions in the deployment of a particular type of microfluidic device (SAR) micromixer, we have recently developed a methodology to produce single devices from scratch to testing without the requirement of ancillary laboratory equipment [10,45]. Combining xurography and lamination offered promising advantages over conventional manufacture such as flexibility, short cycle times, and low-cost. However, to deploy micromixers such as POC, it is important to confront the scalability of the process to produce massively and reliably these devices in the field.

In this work, we present a scaled-up version of a novel split-and-recombine (SAR) array design presented before by our group [10,45] and adapted to be an interface for a handheld multichannel micropipette as a step forward to meet the aforementioned criteria. The manufacturability has been assessed for xurography and laser ablation. A benchmark between these manufacturing processes is relevant as they are the commercially available alternatives for processing thin-films without specialized laboratory equipment.

While the mere availability of POC technologies does not automatically ensure their adoption [46,47], the authors of this article believe that the flexibility of rapid fabrication provided by manufacture based on polymeric films along high-throughput capabilities can ease the integration of micromixing as part of more complex and functional diagnosis, treatment, and monitoring systems. For example, micromixing has shown potential as a low-cost sensitivity enhancer for biosensing micro-devices [48–50].

2. Materials and Methods

A Graphtec CE5000–60 (Graphtec America, Irvine, CA, USA) high precision cutting plotter and a Telesis EV25DS (Telesis Technologies, Circleville, OH, USA) marking system were used to assess xurography and laser ablation. Standard commercial Arlon vinyl sheets (polyvinyl chloride; Arlon Graphics, Placentia, CA, USA) were used for the manufacturability assessment (this vendor was selected considering the availability of a worldwide distributor network). Manufacture using xurography was employed on a testing material with three color variations; gray (G_X), orange (O_X), and black (B_X) using a similar methodology introduced by a previous work for a single micromixer device [10,45] and then adapted and employed on materials of the same batch for gray (G_L), orange (O_L), and black (B_L) using ablation. The procedures can be summarized as follows: a microfluidic device can be developed by patterning and stacking four layers of materials. A PMMA substrate (Layer 0) provides the mechanical stiffness required for handling the device, a vinyl pattern (Layer 1) forms the flow cell walls, and an acetate sheet (Layer 2) is employed to provide an enclosure ceiling to the device and provide additional structural support to the (Layer 4) vinyl sheet that seals the device and delimits the inlets and outlets. Tables 2 and 3 and Sections 2.1 and 2.2 describe more thoroughly the setup conditions.

The implementation of the rapid fabrication methodology studied in this article was assessed in the manufacturing process of a passive micromixing following a procedure presented in previous work [10] and expanded for an array configuration. The implemented design is based on previous research [10,40,45] on an unbalanced split and recombine micromixer. Figure 1a,b shows our proposed testing device setup. A standard 96 microwell microplate (Sigma-Aldrich, Saint Louis, MO, USA) or microtiter is used to store samples prior processing. A standard multichannel micropipette (Thermo Fischer, Waltham, MA, USA) was used to collect the samples from the microwells to introduce them in a four channel microdevice (I to IV). Figure 1b describes the array configuration: each of the channels is integrated by two inputs (I_A, I_B, I_C, etc.), a six phase asymmetric split, and a recombine micromixer; whereas the input main channel (w_{input} = 1500 μm) is divided subsequently into a main subchannel ($w_1 = w_3 = w_5 = w_7$ = 1000 μm) and a lesser subchannel ($w_2 = w_4 = w_6 = w_8 = w_{10} = w_{12}$ = 500 μm). The inputs and outputs of the device ($\varnothing_{input} = \varnothing_{output}$) are equally spaced ($x_d = x_m$ = 9 mm) and share the same dimension pipette to pipette distance. The overall size of the micromixer array is 92 mm × 86.5 mm.

2.1. Xurography Setup

A Graphtec CE5000-60 (Graphtec America, Irvine, CA, USA) high precision cutting plotter was used to pattern 4500 CalPlus vinyl sheet (Layer 1, Arlon Graphics LLC, Placentia, CA, USA). This equipment has a list price around $2000 USD.

Figure 1b describes the four layers composing the micromixing arrays. Commercial acetate sheets transparency foils were used to pattern the intermediate layer (Layer 2; 21.59 × 27.94 cm). The thickness of the acetate sheets was found to be variable among the stock and was examined using confocal microscopy prior experimentation. Devices were sealed using a CalPlus 5000 transparent polymeric film (Layer 3, Arlon Graphics LLC, Placentia, CA, USA) to the cutting machine software (Graphtec Design Studio, Graphtec America, Irvine, CA, USA) for patterning Layers 1, 2, and 3. The cutting tool used for this work was a standard carbide cutting tool model CB09U with a cutting diameter of 0.9 mm and a cutting angle of 45°. Tables 2 and 3 describe the details of the setup conditions for patterning the materials.

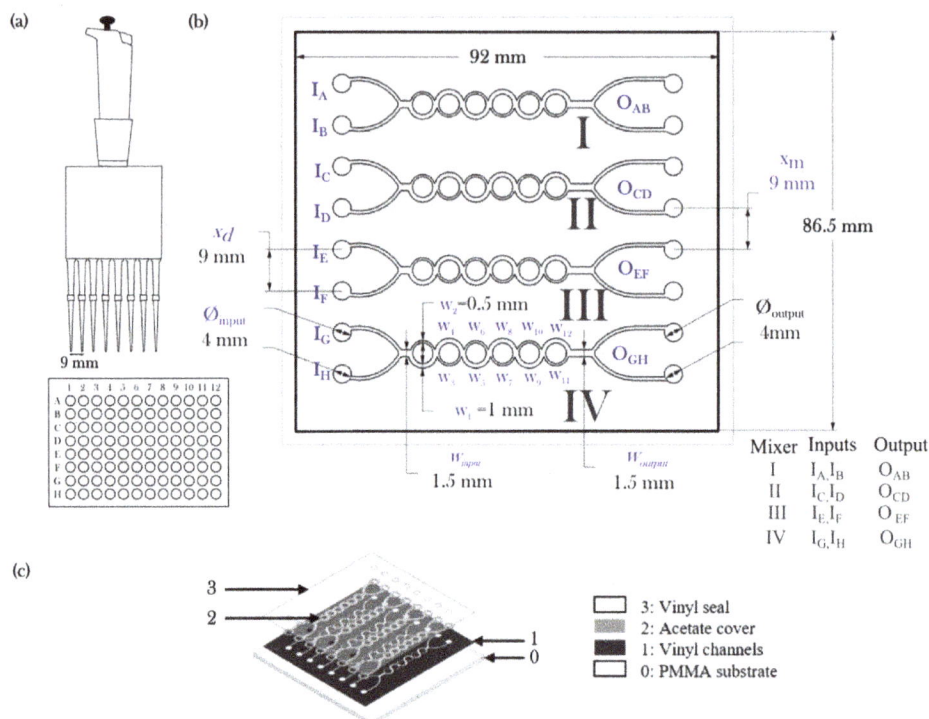

Figure 1. (**a**) Schematics of standardized handheld multichannel micropipette and microwell plate; (**b**) four-channel asymmetric split-and-recombine (ASAR) micromixer (two inputs and two outputs) microdevice setup used to evaluate xurography and laser ablation; (**c**) conforming layers of a microdevice.

Table 2. Setup conditions for laser ablation and xurography.

Setup	Manufacture Technology	Patterning Mechanism	Patterning Conditions	Testing Material
G_X,O_X,B_X	Xurography: Graphtec CE5000-60	Blade CB09U (45°)	Fload $\approx$ 0.8 N, Number of passes = 1	Gray, Orange, Black 4500 CalPlus
G_L,O_L,F_L	Laser ablation: Telesis EV25DS	Q-switched Nd: YVO$_4$ laser	Mark speed = 500 mm/min, Frequency = 10 kHz, Laser power = 22.5 W, Pass number = 10	Gray, Orange, Black 4500 CalPlus

Table 3. Machine setup conditions for the laser ablation process.

Condition	Specification
Laser type	Class 4, fiber-coupled, diode-pumped, Q-switched Nd: YVO$_4$
Wavelength	1064 nm
Mode	TEM_00
Cooling system	Air-cooled
Galvanometer repeatibility	<22 micro radian
Field resolution	16 bit (65,535 data points)
Marking field size (420 mm lens)	290 × 290 mm

2.2. Laser Ablation Setup

Untreated Arlon vinyl sheets (100 mm × 100 mm) with the same features mentioned in Table 2 were adhered to commercial PMMA sheets (Layer 0). A fiber-coupled diode end pumped Q-switched Nd: YVO$_4$ laser was used for patterning the micromixing array. Table 2 presents the main features of the Telesis equipment employed in the experimental trials. This machine is an industrial laser engraver with a list price around $46,000 USD, including hardware and software. The laser beam was focused on vinyl sheets using a 420 mm focal lens and a galvo-mechanism resulting in a theoretical spot size of 127 μm. Trajectories were defined according to a DXF file performed in Autocad (Autodesk, Mill Valley, CA, USA), and marking parameters were established on Merlin II LS software (Telesis, Circleville, OH, USA). The assembly process was made as follows; once the laser ablated the vinyl material (Layer 1) from the PMMA substrate (Layer 0), a tweezer was used to separate the channel geometry from the PMMA substrate (Layer 0). Layer 2 and Layer 3 were added subsequently manually in a similar manner as the samples processed with xurography.

2.3. Array Characterization

Distilled water and food colorant (blue) were used for the visual inspection of the microdevices for the detection of leaks. Microdevice characterization was carried out with an SV8 Zeiss stereoscopic microscope (Carl Zeiss Microscopy, Cambridge, UK) to measure the microdevice dimensions (channel widths and lengths). To evaluate the quality of the patterning process among the conditions, micromixing array elements and features were measured with two replicas. In total, 192 measurements were made to assess the patterning process quantitatively. To compare the dimensional errors between xurography and laser ablation, the absolute average dimensional error was employed, which comprises the percentage error of the undercuttings and overcuttings by considering the mean of the absolute values of the differences between the experimental and intended feature sizes expressed in a percentage.

A confocal microscope Axio CSM 700 (Carl Zeiss Microscopy) was employed to determine the depth of the microdevices during the different steps of the lamination process.

2.4. Micromixing Characterization

The examination and quantification of the micromixing performance were assessed using a custom made machine vision system. Dilutions of a food color additive and distilled water samples were prepared and placed in an intercalated order in a microwell and then transferred into the device by releasing 40 μL droplets of the samples over the inputs of the channel using a multichannel micropipette. Distilled water and a standard compressed air duster (Office Depot, Boca Raton, FL, USA) were used to wash the devices before and after the assessment. The food color additive was composed of water, glycerin, and corn syrup.

To quantify the mixing behavior, the variance of the liquid species in the micromixer (σ) was calculated based on the Danckwerts' segregation intensity index [51]. The variance of the mass fraction of the mixture in a cross-section (σ) that is normal to the flow direction is defined as follows:

$$\sigma = \sqrt{\frac{1}{N}\left(I_{ni} - \overline{I_{nm}}\right)^2},$$ (1)

where N is the number of sampling points inside the cross section, I_{ni} is the normalized intensity value at pixel I, and $\overline{I_{nm}}$ is the normalized mean intensity value at the target area. To quantitatively analyze the numerical mixing performance of the micromixer, the mixing index (M) at a cross-sectional plane is defined as:

$$M = 1 - \sqrt{\frac{\sigma^2}{\sigma_{max}^2}}$$ (2)

where the mixing efficiency ranges from 0.00 (0% mixing) to 1.00 (100% mixing). The maximum variance σ_{max} represents a completely unmixed condition ($\sigma_{max} = 0.5$). The target areas (138 $\times$ 40 pixels) were defined at each output of the channels of the micromixers to extract intensity mean and variance values. Considering the non-uniform flow derived from the tensional passive drive flow produced by the multi-channel micropipette, the intensity was normalized with the concentration of the blue dye.

A C930 Logitech digital camera (Logitech International, Lausanne, Switzerland) with 1920 $\times$ 1080 pixels and a 90-degree field of view was employed to capture the images for processing. Mixing quantification was done using a custom made software using Visual Basic (Microsoft Corporation, Redmond, WA, USA) and Aforge.NET, an open source framework for researchers in Computer Vision and Artificial Intelligence [52].

3. Results and Discussion

Assessment of the rapid manufacturing techniques showed that both manufacturing techniques were tested for different materials and setups (see Table 2 for details) to evaluate the absolute deviational error along the micromixer array elements (from I to IV) of each of the features. Figures 2 and 3 describe the average absolute errors for laser ablation and xurography, respectively. The error is comprised by undercutting and overcutting for each of the setup conditions during the patterning of the gray (G_L,G_X), orange (O_L,O_X), and black (B_L,B_X) materials. The measure of the depth of the conforming Layer 1 and hence the depth of the micromixer (d_m) were consistently found to be approximately 100 µm, regardless of the color of the material. The magnitude of the deviational errors varied greatly depending on the type of feature. For example, inputs and outputs ($\varnothing_{input}$ and $\varnothing_{output}$) were processed consistently with errors below 5%, the wider main channels (w_{input} and w_{output}) and the main subchannels showed higher errors (from 5% to 12%), and the most finessed regions of the device such as the minor subchannels (w_1, w_3, w_5, w_7, w_{11}) displayed the highest deviational errors with values up to 42%. It should be mentioned that, for all the patterning setups, the differences among channels was low, suggesting that the precision of the channel is suitable for further scaling up of the number of micromixing channels. Among all the setups, the G_L configuration (gray vinyl under laser ablation) showed the lowest device to device variability (standard deviation) among the measurements.

Technological considerations arose during the assessment of the laser ablation experimentation as follows. The results indicate an absolute dimensional error that was larger in lesser channels, which is explained by heat affected zones where thermal diffusivity is promoted in narrow zones (e.g., 500 µm subchannels) and consequently cut widths with greater variability. Also, the differences among orange, black, and gray setups are explained due to the coefficient absorption of Nd: YVO$_4$ by differences among the optical properties of the materials. The aforementioned condition explains the steep differences in quality performance among the laser ablation setups (Figure 2) and the similar overall performance among xurography setups. Moreover, the use of laser engraving or laser cutting systems in industrial settings can lead to a regular exposure of particles generated during material processing. For the assessment, we had considered a report on the risks of exposure of polyvinyl chloride in an industrial setting [53] to carry out an evaluation in an enclosed chamber. While it is still necessary to determine the magnitude of the risks for human health and security, the circumstance adds a limitation layer for the utilization of laser ablation, especially aiming towards high-throughput performance for the POC setting.

The absolute dimensional error for then xurography based manufacturing process (see Figure 3) was shown to be below 5% for the circular patterns for the inputs and the outputs and higher among the features with finer and more complex details such as the the minor subchannels (w_2, w_3, w_6, w_7, w_{10}, w_{11}). Differences in performance among the array elements can be attributed to uneven conditions of the material properties of the surface, errors in positioning derived from the moving mechanical parts, and the gradual loss of sharpness of the blade.

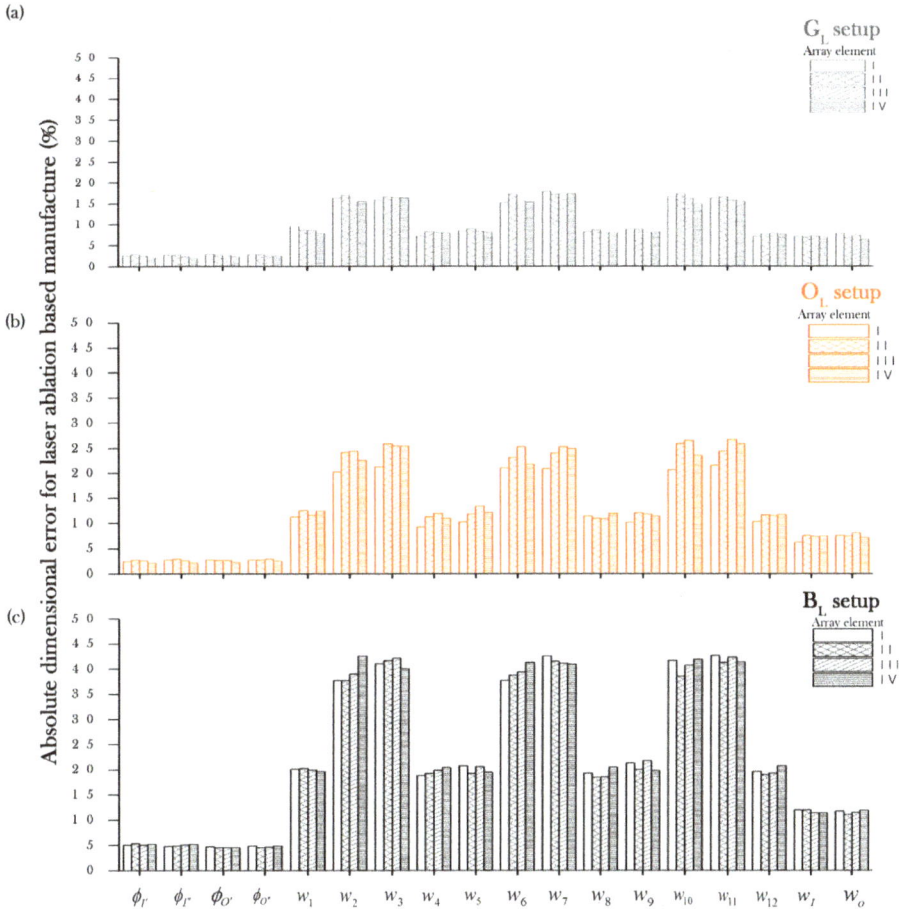

Figure 2. Absolute deviation of the geometrical features of an asymmetric split-and-recombine (ASAR) micromixer for a laser ablation-based manufacture process for (**a**) gray vinyl (G_L), (**b**) orange vinyl (O_L) and (**c**) black vinyl (B_L).

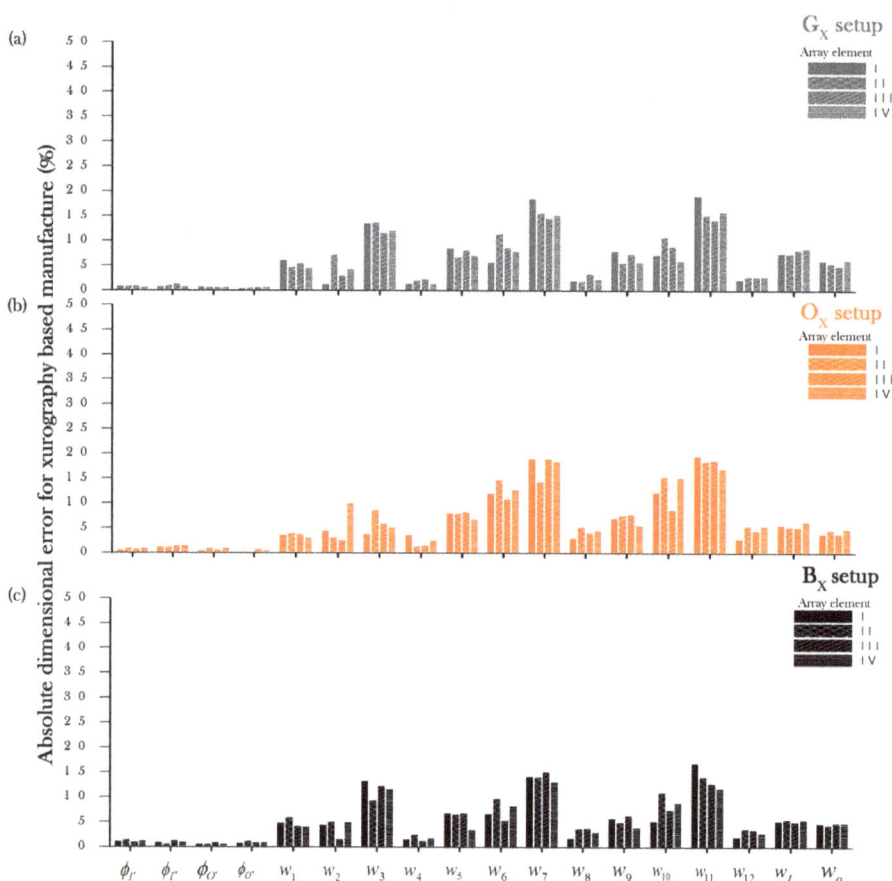

Figure 3. Absolute deviation of the geometrical features of an asymmetric split-and-recombine (ASAR) micromixer for a xurography-based manufacture process for (**a**) gray vinyl (G_X), (**b**) orange vinyl (O_X) and (**c**) black vinyl (B_X).

The overall performance of laser ablation and xurography is shown in Figure 4a. Setup B_X (black vinyl under xurography) showed among all the setups the highest quality, with only 5.41% absolute deviational error. For the laser ablation, we were able to manufacture the array of devices within comparable values from previous research for the G_L setup; however the quality of the orange O_X and B_X setups suggest that improving the quality of Layer 1 of the device requires optimization using some of the other laser ablation parameters.

To confirm the B_X setup as a suitable manufacture methodology for a multi-channel ASAR micromixer array, an evaluation of the mixing performance was done following the procedure described in Section 2.3. Figure 4b shows pictures of 0, 10, 40, 100, and 160 s after the sample was introduced on the inlets simultaneously using the multichannel micropipette. The micromixing efficiency was estimated at the output of the channels. A video of the experimental test can be found in a data archiving service and as Supplementary Material (video S1: Micromixing test of a four channel split and recombine array) [54].

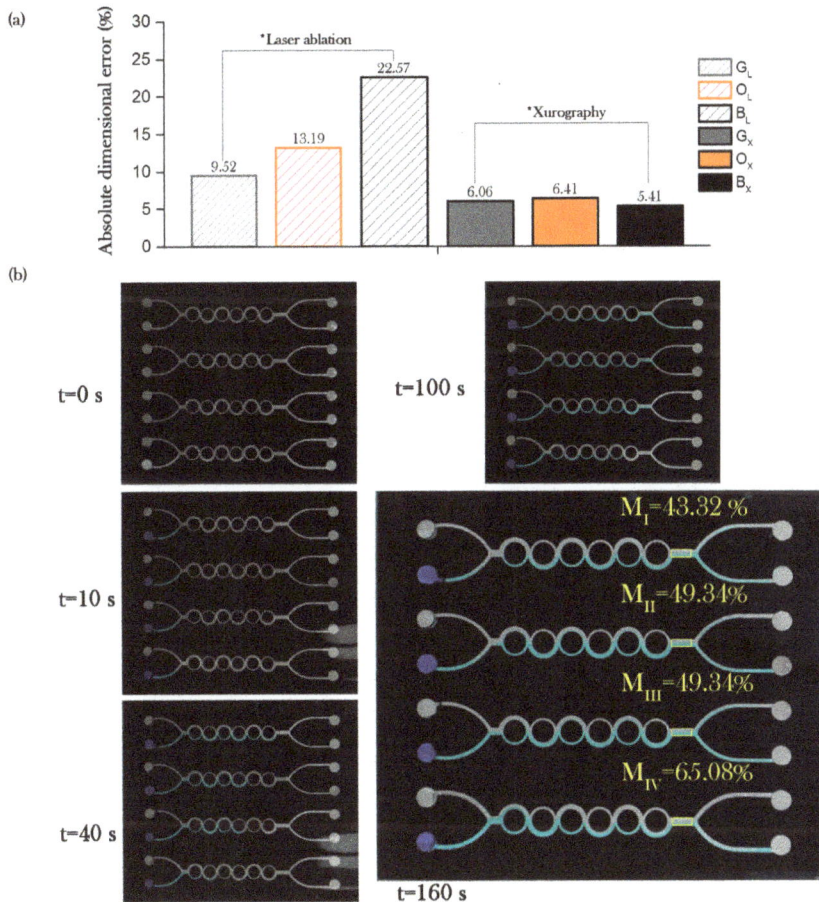

Figure 4. (a) Overall absolute deviational error for laser ablation and xurography based manufacture; (b) example of a four-channel device mixing after *t* = 0, 10, 40, 100, and 160 s of passive driven flow, mixing efficiencies are indicated in yellow for each of the array elements.

The Reynolds number is conventionally used to characterize the fluidic behavior in microdevices and is defined as the ratio of inertial to viscous forces. Equation (3) represents the Reynolds number (*Re*), defined as:

$$Re = \frac{\text{inertial force}}{\text{viscous force}} = \frac{\rho U D_{\text{H}}}{\mu}. \tag{3}$$

where μ is the viscosity(Pa·s), ρ is the fluid density (kg·m^{-3}), U is the average velocity of the flow (m·s^{-1}), and D_{H} is the hydraulic diameter of channel (m), which is defined in Equation (4) as:

$$D_{\text{H}} = \frac{4A}{P} = \frac{4wd_{\text{m}}}{2w + 2d_{\text{m}}}, \tag{4}$$

where A and P are, respectively, the area and the wetted perimeter of the cross-section, which is given by the micromixer width ($w = w_{\text{input}}$) and depth of the channel (d_{m}).

The water properties at 20 °C density (ρ) were considered in 9.998 × 10^2 kg·m^{-3} and the dynamic viscosity (σ) in 1.01 × 10^{-3} kg·m^{-1}·s^{-1} [32]. The mixing efficiency was evaluated using the

Danckwerts' segregation intensity index (see Section 2.2) 160 s after the samples were introduced in the microdevice. To estimate the Reynolds number and hence determine the flow regime in which each of the micromixers array elements is operating under the flow velocity, U was estimated by measuring the time required for the dye to reach the output (interrogation window).

Mixing evaluation windows are indicated in yellow at the outputs of the mixing array micromixers elements at $t = 160$ s. The Reynolds number was estimated ranging from 0.07 to 0.13, with corresponding mixing efficiency varying from 43.32% up to 65.08% (see Table 4). While the lateral dimensions of the microfluidic devices are larger than other in-plane micromixers reported in the literature (see Table 1), the flow regime remains laminar considering the device remains operating below 2300. To determine the source of the differences among the flow velocities between the array elements requires further experimental work. Differences in the average flow velocity are prone to arise due to various factors including the geometrical features of the channel produced by errors during the patterning and assembly of the conforming layers, non-uniform dispense of the liquid samples during pipetting, and clogging caused by debris or bubbles located inside the flow chamber during experimentation.

Table 4. Flow conditions and mixing efficiency of a four channel micromixer device manufactured using xurography.

Mixing Array Element	Average Flow Velocity (U)	Reynolds Number (Re)	Mixing Efficiency (M)
I	0.7 mm/s	0.13	43.32%
II	0.5 mm/s	0.09	49.34%
III	0.47 mm/s	0.08	49.34%
IV	0.38 mm/s	0.07	65.08%

4. Conclusions and Future Work

- The dimensional accuracy of xurography was shown to be better for xurography than laser ablation for the ASAR micromixing array. Compared to xurography, the deployment of the laser ablation as a manufacturing tool in the POC setting underwent several disadvantages such as the requirement to adjust the setup parameters regarding the optical properties of the material and the additional health and security considerations for the laser processing of materials.

- Assessments of both the rapid manufacture technologies were successfully employed to produce low-cost microfluidic device arrays with deviational errors below 10% under certain setup conditions for xurography and laser ablation.

- Small differences in the dimensional errors among different ASAR micromixer members suggests that it is possible to scale-up further the size of the array.

- The proposed four element micromixer array design was successfully coupled with a standardized multichannel micropipette for micromixing simultaneously eight samples of dye with mixing performance up to 65%.

- The proposed design interfaces standardized dispensing (handheld micropipette) and sampling (microplate well) equipment.

- In the future, it is necessary to validate the mixing performance of the micromixing devices under different conditions (materials, geometries, instrumentation setup). Additional research is also required to determine factors affecting the systematic dimensional errors found in certain components of the micromixing device.

Supplementary Materials: The following is available online at www.mdpi.com/2072-666X/8/5/144/s1, Video S1: Micromixing test of a four channel split and recombine array.

Acknowledgments: The CONACyT Basic Scientific Research Grant No. 242634 has supported this work. The Research Group of Advanced Manufacturing at Tecnológico de Monterrey provided additional funding. J. Israel Martínez-López thank Alex Elías Zúñiga for easing access to equipment and information.

Author Contributions: J.I.M.L. conceived and designed the micromixer and evaluated the mixing performance data, he was also responsible of writing major extents of the article. H.A.B. performed the manufacturing processes and collaborated with the mixing characterization. E.G.-L. designed the setup conditions for the laser ablation process. J.I.M.L. and H.R.S. contributed insights regarding the assessment analysis. C.A.R. contributed to the analysis of data and aided in the selection, operation, and performance of the manufacturing technologies.

Conflicts of Interest: The authors declare no conflict of interest.

Abbreviations

The following abbreviations are used in this manuscript:

AIDS	Acquired Immune Deficiency Syndrome
ASAR	Asymmetric split and recombination
CAD	Computer aided design
COC	Cyclic olefin copolymer
COP	Cyclic olefin polymer
DXF	Drawing Interchange Format
EP	Electrophoresis
FDM	Fused deposition modeling
SAR	Split and recombine
SGM	Slanted grooved mixer
SHM	Staggered herringbone mixer
PC	Polycarbonate
PET	Polyethylene terephthalate
PETG	Polyethylene terephthalate glycol
PDMS	Polydimethylsiloxane
PEEK	Polyether ether ketone
PMMA	Polymethyl methacrylate
POC	Point-of-Care
PVC	Polyvinyl chloride
RGB	Red-green-blue
SL	Stereolitography
TB	Tuberculosis
WHO	World Health Organization

References

1. Bissonnette, L.; Bergeron, M.G. Diagnosing infections—Current and anticipated technologies for point-of-care diagnostics and home-based testing. *Clin. Microbiol. Infect.* **2010**, *16*, 1044–1053. [CrossRef] [PubMed]
2. Mabey, D.; Peeling, R.W.; Ustianowski, A.; Perkins, M.D. Diagnostics for the developing world. *Nat. Rev. Microbiol.* **2004**, *2*, 231–240. [CrossRef] [PubMed]
3. Chin, C.D.; Linder, V.; Sia, S.K. Lab-on-a-chip devices for global health: Past studies and future opportunities. *Lab Chip* **2006**, *7*, 41–57. [CrossRef] [PubMed]
4. Bell, D.; Peeling, R.W. Evaluation of rapid diagnostic tests: Malaria. *Nat. Rev. Microbiol.* **2006**, *4*, S34–S38. [CrossRef] [PubMed]
5. Banoo, S.; Bell, D.; Bossuyt, P.; Herring, A.; Mabey, D.; Poole, F.; Smith, P.G.; Sriram, N.; Wongsrichanalai, C.; Linke, R.; et al. Evaluation of diagnostic tests for infectious diseases: general principles. *Nat. Rev. Microbiol.* **2008**, *8*, S16–S28. [CrossRef] [PubMed]
6. Natarajan, S.; Chang-Yen, D.A.; Gale, B.K. Large-area, high-aspect-ratio SU-8 molds for the fabrication of PDMS microfluidic devices. *J. Micromech. Microeng.* **2008**, *18*, 045021. [CrossRef]
7. McDonald, J.C.; Whitesides, G.M. Poly(dimethylsiloxane) as a material for fabricating microfluidic devices. *Acc. Chem. Res.* **2002**, *35*, 491–499. [CrossRef] [PubMed]
8. Ren, K.; Zhou, J.; Wu, H. Materials for microfluidic chip fabrication. *Acc. Chem. Res.* **2013**, *46*, 2396–2406. [CrossRef] [PubMed]

9.   Mogi, K.; Sugii, Y.; Yamamoto, T.; Fujii, T. Rapid fabrication technique of nano/microfluidic device with high mechanical stability utilizing two-step soft lithography. *Sens. Actuators B Chem.* **2014**, *201*, 407–412. [CrossRef]

10. Martínez-López, J.I.; Mojica, M.; Rodríguez, C.A.; Siller, H.R. Xurography as a rapid fabrication alternative for point-of-care devices: Assessment of passive micromixers. *Sensors* **2016**, *16*, 705. [CrossRef] [PubMed]

11. Ventola, C.L. Medical applications for 3D printing: Current and projected uses. *Pharm. Ther.* **2014**, *39*, 704–711.

12. Yazdi, A.A.; Popma, A.; Wong, W.; Nguyen, T.; Pan, Y.; Xu, J. 3D printing: An emerging tool for novel microfluidics and lab-on-a-chip applications. *Microfluid. Nanofluid.* **2016**, *20*, 50. [CrossRef]

13. Vig, A.L.; Mäkelä, T.; Majander, P.; Lambertini, V.; Ahopelto, J.; Kristensen, A. Roll-to-roll fabricated lab-on-a-chip devices. *J. Micromech. Microeng.* **2011**, *21*, 035006. [CrossRef]

14. Deng, Y.; Yi, P.; Peng, L.; Lai, X.; Lin, Z. Flow behavior of polymers during the roll-to-roll hot embossing process. *J. Micromech. Microeng.* **2015**, *25*, 065004. [CrossRef]

15. Liedert, R.; Amundsen, L.K.; Hokkanen, A.; Mäki, M.; Aittakorpi, A.; Pakanen, M.; Scherer, J.R.; Mathies, R.A.; Kurkinen, M.; Uusitalo, S.; et al. Disposable roll-to-roll hot embossed electrophoresis chip for detection of antibiotic resistance gene mecA in bacteria. *Lab Chip* **2011**, *12*, 333–339. [CrossRef] [PubMed]

16. Bartholomeusz, D.A.; Boutte, R.W. Xurography: Rapid prototyping of microstructures using a cutting plotter. *Microelectromech. Syst. J.* **2006**, *14*, 1364–1374. [CrossRef]

17. Treise, I.; Fortner, N.; Shapiro, B.; Hightower, A. Efficient energy based modeling and experimental validation of liquid filling in planar micro-fluidic components and networks. *Lab Chip* **2005**, *5*, 285–297. [CrossRef] [PubMed]

18. Silhouette America. Available online: https://www.silhouetteamerica.com/ (accessed on 14 April 2017).

19. Cricut. Available online: https://home.cricut.com/ (accessed on 14 April 2017).

20. Huang, H.; Guo, Z. Ultra-short pulsed laser PDMS thin-layer separation and micro-fabrication. *J. Micromech. Microeng.* **2009**, *19*, 055007. [CrossRef]

21. Liao, Y.; Ju, Y.; Zhang, L.; He, F.; Zhang, Q.; Shen, Y.; Chen, D.; Cheng, Y.; Xu, Z.; Sugioka, K. Three-dimensional microfluidic channel with arbitrary length and configuration fabricated inside glass by femtosecond laser direct writing. *Opt. Lett.* **2010**, *35*, 3225–3227. [CrossRef] [PubMed]

22. Ben-Yakar, A.; Byer, R.L.; Harkin, A.; Ashmore, J.; Stone, H.A.; Shen, M.; Mazur, E. Morphology of femtosecond-laser-ablated borosilicate glass surfaces. *Appl. Phys. Lett.* **2003**, *83*, 3030–3032. [CrossRef]

23. McCann, R.; Bagga, K.; Groarke, R.; Stalcup, A.; Vázquez, M.; Brabazon, D. Microchannel fabrication on cyclic olefin polymer substrates via 1064 nm Nd:YAG laser ablation. *Appl. Surf. Sci.* **2016**, *387*, 603–608. [CrossRef]

24. Grzybowski, B.A.; Haag, R.; Bowden, N.; Whitesides, G.M. Generation of micrometer-sized patterns for microanalytical applications using a laser direct-write method and microcontact printing. *Anal. Chem.* **1998**, *70*, 4645–4652. [CrossRef]

25. Hong, T.-F.; Ju, W.-J.; Wu, M.-C.; Tai, C.-H.; Tsai, C.-H.; Fu, L.-M. Rapid prototyping of PMMA microfluidic chips utilizing a CO_2 laser. *Microfluid. Nanofluid.* **2010**, *9*, 1125–1133. [CrossRef]

26. Malek, C.G.K. Laser processing for bio-microfluidics applications (part II). *Anal. Bioanal. Chem.* **2006**, *385*, 1362–1369. [CrossRef] [PubMed]

27. Pinecone Robotics. Available online: http://www.p-robots.com (accessed on 14 April 2017).

28. Mr Beam-Laser Cutting for Everybody. Available online: http://mr-beam.org/#tech (accessed on 14 April 2017).

29. Nguyen, N.-T.; Wu, Z. Micromixers—A review. *J. Micromech. Microeng.* **2005**, *15*, R1–R16. [CrossRef]

30. Su, Y.; Chen, G.; Yuan, Q. Ideal micromixing performance in packed microchannels. *Chem. Eng. Sci.* **2011**, *66*, 2912–2919. [CrossRef]

31. Schönfeld, F.; Hardt, S. Simulation of helical flows in microchannels. *AIChE J.* **2004**, *50*, 771–778. [CrossRef]

32. Du, Y.; Zhang, Z.; Yim, C.; Lin, M.; Cao, X. Evaluation of floor-grooved micromixers using concentration-channel length profiles. *Micromachines* **2010**, *1*, 19–33. [CrossRef]

33. Stroock, A.D.; Dertinger, S.K.W.; Ajdari, A.; Mezić, I.; Stone, H.A.; Whitesides, G.M. Chaotic mixer for microchannels. *Science* **2002**, *295*, 647–651. [CrossRef] [PubMed]

34. Chung, C.K.; Shih, T.R. Effect of geometry on fluid mixing of the rhombic micromixers. *Microfluid. Nanofluid.* **2007**, *4*, 419–425. [CrossRef]

35. Hossain, S.; Kim, K.-Y. Mixing analysis of passive micromixer with unbalanced three-split rhombic sub-channels. *Micromachines* **2014**, *5*, 913–928. [CrossRef]

36. Hong, C.-C.; Choi, J.-W.; Ahn, C.H. A novel in-plane passive micromixer using coanda effect. In *Micro Total Analysis Systems 2001*; Ramsey, J.M., van den Berg, A., Eds.; Springer: Dordrecht, The Netherlands, 2001; pp. 31–33.

37. Hong, C.-C.; Choi, J.-W.; Ahn, C.H. A novel in-plane passive microfluidic mixer with modified Tesla structures. *Lab Chip* **2004**, *4*, 109–113. [CrossRef] [PubMed]

38. Hossain, S.; Ansari, M.A.; Husain, A.; Kim, K.-Y. Analysis and optimization of a micromixer with a modified Tesla structure. *Chem. Eng. J.* **2010**, *158*, 305–314. [CrossRef]

39. Sudarsan, A.P.; Ugaz, V.M. Fluid mixing in planar spiral microchannels. *Lab Chip* **2006**, *6*, 74–82. [CrossRef] [PubMed]

40. Ansari, M.A.; Kim, K.-Y.; Anwar, K.; Kim, S.M. A novel passive micromixer based on unbalanced splits and collisions of fluid streams. *J. Micromech. Microeng.* **2010**, *20*, 055007. [CrossRef]

41. Ansari, M.A.; Kim, K.-Y. Mixing performance of unbalanced split and recombine micomixers with circular and rhombic sub-channels. *Chem. Eng. J.* **2010**, *162*, 760–767. [CrossRef]

42. Chung, C.K.; Shih, T.R.; Wu, B.H.; Chang, C.K. Design and mixing efficiency of rhombic micromixer with flat angles. *Microsyst. Technol.* **2009**, *16*, 1595–1600. [CrossRef]

43. Scherr, T.; Quitadamo, C.; Tesvich, P.; Park, D.S.-W.; Tiersch, T.; Hayes, D.; Choi, J.-W.; Nandakumar, K.; Monroe, W.T. A planar microfluidic mixer based on logarithmic spirals. *J. Micromech. Microeng.* **2012**, *22*, 055019. [CrossRef] [PubMed]

44. Li, J.; Xia, G.; Li, Y. Numerical and experimental analyses of planar asymmetric split-and-recombine micromixer with dislocation sub-channels. *J. Chem. Technol. Biotechnol.* **2013**, *88*, 1757–1765. [CrossRef]

45. Martínez-López, J.I.; Mojica, M.; Betancourt, H.A.; Rodríguez, C.A.; Siller, H.R. Xurography and Lamination for Manufacturing Point-of-Care (POC) Micromixers. In Proceedings of the 4M/IWMF2016 Conference, Lyngby, Denmark, 13–15 September 2016; Research Publishing Services: Kgs. Lyngby, Denmark, 2016; pp. 227–230.

46. Pai, N.P.; Vadnais, C.; Denkinger, C.; Engel, N.; Pai, M. Point-of-care testing for infectious diseases: Diversity, complexity, and barriers in low-and middle-income countries. *PLoS Med.* **2012**, *9*, e1001306. [CrossRef] [PubMed]

47. Engel, N.; Ganesh, G.; Patil, M.; Yellappa, V.; Pai, N.P.; Vadnais, C.; Pai, M. Barriers to point-of-care testing in India: Results from qualitative research across different settings, users and major diseases. *PLoS ONE* **2015**, *10*, e0135112. [CrossRef] [PubMed]

48. Martínez-López, J.I. *Biosensing Enhancement of a SPR Imaging System Through Micromixing Structures*; Tecnologico de Monterrey: Monterrey, Nuevo León, Mexico, 2013.

49. Lynn, N.S.; Martínez-López, J.-I.; Bocková, M.; Adam, P.; Coello, V.; Siller, H.R.; Homola, J. Biosensing enhancement using passive mixing structures for microarray-based sensors. *Biosens. Bioelectron.* **2014**, *54*, 506–514. [CrossRef] [PubMed]

50. Ko, Y.-J.; Maeng, J.-H.; Ahn, Y.; Hwang, S.Y. DNA ligation using a disposable microfluidic device combined with a micromixer and microchannel reactor. *Sens. Actuators B Chem.* **2011**, *157*, 735–741. [CrossRef]

51. Danckwerts, P.V. The definition and measurement of some characteristics of mixtures. *Appl. Sci. Res. Sect. A* **1952**, *3*, 279–296.

52. AForge.NET:: Computer Vision, Artificial Intelligence, Robotics. Available online: http://www.aforgenet.com/ (accessed on 28 February 2017).

53. Langard, S.; Rosenberg, J.; Andersen, A.; Heldaas, S. Incidence of cancer among workers exposed to vinyl chloride in polyvinyl chloride manufacture. *Occup. Environ. Med.* **2000**, *57*, 65–68. [CrossRef] [PubMed]

54. López, J.I.M.; Cervantes, H.A.B.; Garcia-Lopez, E.; Rodriguez, C.; Siller, H.R. Micromixing Test of a Four Channel Split and Recombine Array. Available online: https://doi.org/10.6084/m9.figshare.4704769.v1 (accessed on 26 April 2017).

micromachines

MDPI

Article

Ultrasonic-Assisted Incremental Microforming of Thin Shell Pyramids of Metallic Foil

Toshiyuki Obikawa *,† and **Mamoru Hayashi** ‡

Institute of Industrial Science, The University of Tokyo, Tokyo 153-8505, Japan; mamolu@ba2.so-net.ne.jp
* Correspondence: obikawa@mail.dendai.ac.jp; Tel.: +81-3-5284-5902
† Current address: Center for Manufacturing Science and Technology, Tokyo Denki University, Tokyo 120-8551, Japan.
‡ Current address: Nidec Center for Industrial Science, Nidec Corporation, Kawasaki 212-0032, Japan.

Academic Editor: Guido Tosello
Received: 28 February 2017; Accepted: 25 April 2017; Published: 3 May 2017

Abstract: Single point incremental forming is used for rapid prototyping of sheet metal parts. This forming technology was applied to the fabrication of thin shell micropyramids of aluminum, stainless steel, and titanium foils. A single point tool used had a tip radius of 0.1 mm or 0.01 mm. An ultrasonic spindle with axial vibration was implemented for improving the shape accuracy of micropyramids formed on 5–12 micrometers-thick aluminum, stainless steel, and titanium foils. The formability was also investigated by comparing the forming limits of micropyramids of aluminum foil formed with and without ultrasonic vibration. The shapes of pyramids incrementally formed were truncated pyramids, twisted pyramids, stepwise pyramids, and star pyramids about 1 mm in size. A much smaller truncated pyramid was formed only for titanium foil for qualitative investigation of the size reduction on forming accuracy. It was found that the ultrasonic vibration improved the shape accuracy of the formed pyramids. In addition, laser heating increased the forming limit of aluminum foil and it is more effective when both the ultrasonic vibration and laser heating are applied.

Keywords: incremental microforming; ultrasonic spindle; shin shell micropyramid; metallic foil; forming limit; shape accuracy

---

1. Introduction

Functional miniaturized structures of various materials will be widely applied to sensors, filters, biotesters, bioscaffold, etc. in the near future. For this purpose, efficient microprocessing technologies with high accuracy and high surface quality are required to meet the demand for their production. Microcutting [1], micro-laser-machining [2], micro-electric discharge machining [3,4], and microforming [5] have been intensively studied, and recently, micro-additive-manufacturing combined with other microprocessing [6] is more and more important in tailor-made manufacturing or high-mix low-volume production of complex parts.

Incremental forming with a single point tool is regarded as a rapid prototyping process for sheet metal forming because no die is needed and parts of free-form surfaces can be fabricated incrementally [7]. The forming limit is much larger for this technology than for other forming methods [8]. Therefore, this technology can be applied to the fabrication of various shapes of sheet metal parts [9]. It also can be applied to microforming of the miniature shell structures of aluminium, gold, and stainless steel foils [10–14]. The preparation of backing plates of various shapes, which are used as a support in ordinary incremental forming, is not easy for incremental forming of microparts. For this reason, no backing plate is needed for the developed microforming technologies. However, the accuracy of microshell structures formed without using a backing plate is not so high, and thus, it should be improved without reducing the forming limit.

In this paper, ultrasonic assisted incremental microforming technology has been developed using an ultrasonic spindle for improving the accuracy of microshell structures of aluminium, stainless steel, and titanium foils. Laser heating was also applied to the ultrasonic assisted incremental microforming for softening and increasing formability of aluminum foil under deformation. The forming limit and forming accuracy of micropyramids were investigated under the conditions of with and without ultrasonic vibration and laser heating. Only for a foil of titanium—a difficult-to-work material—a micropyramid in a size of 283 μm was also formed under the conditions of ultrasonic vibration for qualitatively investigating the size reduction on forming accuracy.

2. Materials and Methods

An incremental microforming machine with an ultrasonic spindle is shown in Figure 1. It is placed on a vibration-free table for insulating external disturbances. It is composed of an x-y table, z stage, ultrasonic spindle with a Langevin ultrasonic transducer, forming tool, blank holder, base, column, and laser source and its optical system. An ordinary motor spindle used in the original and improved forming machines [10–13] has been replaced with an ultrasonic spindle and laser source and its optical system has been newly installed. The ultrasonic spindle causes vibration in the axial direction of the spindle. The vibration frequency is 42.5 kHz and its amplitude is set to be 0.5 μm, the minimum amplitude of the spindle. The x-y table and z-stage are controlled with a personal computer numerically. Their resolutions of motion are 0.01 μm, which is small enough to fabricate a shell structure of the order of one millimeter.

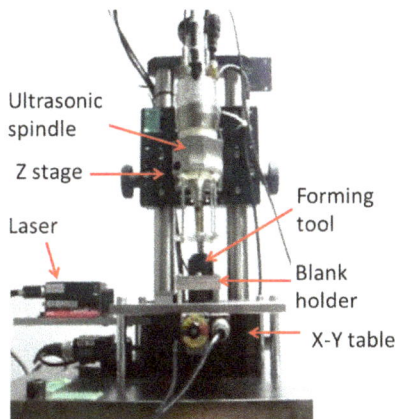

Figure 1. Desktop microforming machine.

Aluminum foil of type 8021 used for the experiments is 6.5 and 12 μm-thick, while stainless steel foil of type 304 is 8 μm-thick and titanium foil is 5 μm-thick. The 8021 foil has a chemical composition listed in Table 1 and a crystal grain size of 5–10 μm [15]. The elongation of metallic foils decreases with decreasing thickness. It is about five percent for the aluminum foils [15] and about one percent for stainless foil [16]. A blank sheet is put on matte side up between the tensioner and O-ring, and then, it was clamped in a blank holder as shown in Figure 2. The holder can apply an appropriate size of tension to a blank by adjusting the sizes of a tensioner and O-ring [10–12]. This method for applying tension to a blank was also adopted in reference [14]. As described elsewhere [10–12] in details, no backing plate supporting a blank is used because miniaturization of a backing plate is not easy.

Table 1. Chemical composition of 8021 aluminum foil [wt %].

Si	Fe	Cu	Mn	Mg	Zn	Others	Al
<0.15	>1.2, <1.7	<0.05	<0.05	<0.05	<0.05	<0.05	Remainder

A single point forming tool of ultra-fine grain cemented carbide with a tip radius of $R = 100$ μm or 10 μm is used in this study (Figure 3). Before starting microforming, z-position of the top surface of a blank was determined accurately by detecting the contact between the tool and blank. n-propyl alcohol is used as a lubricant for avoiding adhesion and abrasion between the forming tool and blank. High speed rotation of the tool generates the hydrodynamic pressure between the rotating tool and blank being formed. Under this condition, the alcohol may penetrate into the interface between them if the contact stress is low.

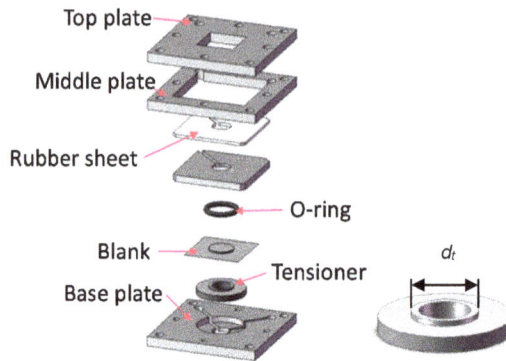

Figure 2. Blank holder and tensioner.

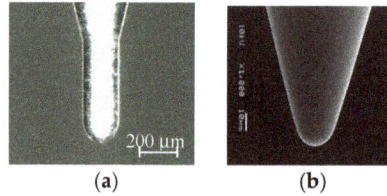

Figure 3. Microtool for incremental microforming: (**a**) $R = 100$ μm; (**b**) $R = 10$ μm.

Incremental microforming process of a triangular pyramid, for example, is shown in Figure 4, where D is the diameter of a circumscribed circle of a base of a pyramid, α is a half apex angle defined as an angle between a lateral edge and a vertical line through an apex, θ is a half apex angle defined as an angle between a triangular lateral face and the vertical line, t is the thickness of a blank, ω is the tool rotational speed, and Δz is the axial feed per a planar tool path. The side length of a triangular tool path on a plane shrinks step by step. The tool rotational speed ω and table speed v_t were set to be 5000 min^{-1} and 200 μm/s, respectively, except that $\omega = 10{,}000$ min^{-1} and $v_t = 150$ μm/s for a titanium micropyramid of $D = 283$ μm. The axial feed Δz was 5 and 12 μm for 6.5 and 12 μm-thick aluminum foils, respectively, 5 μm for 8 μm-thick stainless steel foil, and 2 and 1 μm for titanium pyramids of $D = 1.41$ mm and $D = 283$ μm, respectively. When the forming was conducted under the condition of laser heating, the forming area of the foil was irradiated from the top of a pyramid (from the below of a pyramid) using a semiconductor laser of 130 mW; the laser beam was reflected using a prism under the blank holder.

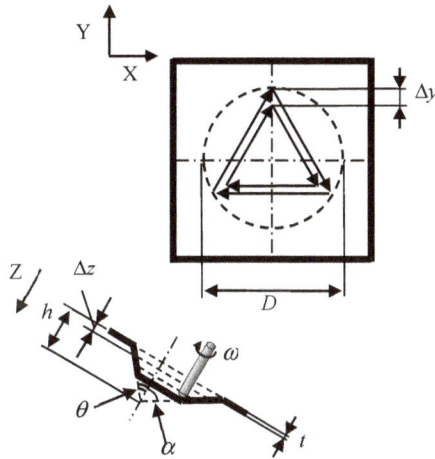

Figure 4. Schematic diagram of tool path and foil deformation during forming of a triangular pyramid.

The triangular lateral faces of a square pyramid elongates from h_0 to h_1 by single point incremental forming as shown in Figure 5. Its elongation e in percent is given by

$$e = 100(h_1/h_0 - 1) = 100(\text{cosec } \theta - 1) \tag{1}$$

while the logarithmic strain ε corresponding to the elongation e is

$$\varepsilon = \ln(h_1/h_0) = \ln(\text{cosec } \theta) \tag{2}$$

Both the elongation and strain increase with decreasing half apex angle. Equations (1) and (2) are applied for truncated pyramids with the same half apex angle θ.

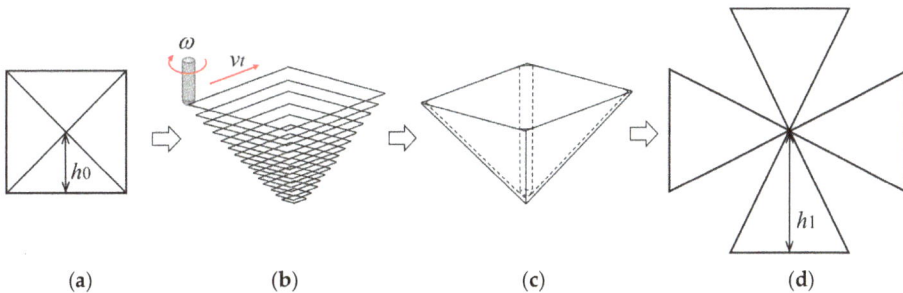

Figure 5. Forming of a square pyramid: (**a**) square blank to be formed; (**b**) tool path; (**c**) formed pyramid; (**d**) development view of a square pyramid.

3. Results and Discussion

The influence of ultrasonic vibration on the forming limit of a truncated pyramid of 6.5 µm-thick aluminum foil is shown in Figure 6, whereas, those of ultrasonic vibration and laser heating on the forming limit is shown in Figure 7. The diameter of a circumscribed circle of a square pyramid base D was 1.41 mm. Forming conditions of with and without ultrasonic vibration are denoted as V and NV, respectively and those of with and without laser heating are denoted as H and NH, respectively. In these figures, the values of half apex angle θ and corresponding strain ε are written in rows above micrographs of the top views of formed pyramids. The value of θ is changed step by step so that the value of ε increases by about 0.05.

θ	39.5°	37.4°
ε	0.45	0.50
NV & NH		
V & NH		 1.0 mm

Figure 6. Forming limits for truncated pyramids with and without ultrasonic vibration (V and NV) and without laser heating (NH).

θ	37.4°	35.2°
ε	0.50	0.55
NV & H		
V & H		
θ	32.9°	30.0°
ε	0.61	0.69
V & H	 Pin hole	 1.0 mm

Figure 7. Forming limits for truncated pyramids with and without ultrasonic vibration (V and NV) and with laser heating (H).

It is seen in Figure 6 that square pyramids of θ = 39.5° were formed without cracks under the conditions of with and without ultrasonic vibration, but cracks grew especially along the lateral edges of pyramids of θ = 37.4° formed under both the conditions. Thus, the forming limit ε_c was 0.45 for the condition of without laser heating and the ultrasonic vibration increased the forming limit marginally. The fact that the forming limit obtained above is less than those obtained in the previous research works [11,12] is partly because n-propyl alcohol is a poor lubricant compared with pure water used in the previous research and partly because the thickness of aluminium foil was reduced from 12 μm to 6.5 μm.

In the case of laser heating, it is found from Figure 7 that ε_c was 0.55 and 0.50 for the conditions of with and without ultrasonic vibration, respectively. This indicates that the laser heating can increase the forming limit of the aluminum foil and it is more effective when the ultrasonic vibration is applied. Because only a pinhole was found on a pyramid of θ = 32.9° under the conditions of with both ultrasonic vibration and laser heating, optimization of forming parameters may increase the forming limit under these conditions. It should be noted that elongation for θ = 35.2°, which is calculated to be 73.4% from Equation (1), is more than 10 times as large as that of about 5% obtained for tensile test.

In addition to the formability, it is confirmed that almost all the cracks appeared on the lateral edges of formed pyramids. This fact is consistent with the results obtained in reference [13], that a crack nucleated on a pyramid edge in the microforming of aluminum foil, whereas it nucleated on a triangular pyramid face in the microforming of stainless steel foil.

The effect of ultrasonic vibration assistance on the shape accuracy in incremental microforming was investigated by forming a twisted pyramid of 6.5 μm-thick aluminum foil and a star pyramid of 12 μm-thick aluminum foil with and without ultrasonic vibration. Diameter D was 1.41 mm for the twisted pyramid and 1.60 mm for the star pyramid. Figure 8 shows the top views of the twisted pyramids incrementally formed. The surfaces and edges of the pyramid made by ultrasonic assisted microforming are much smoother and not wavier than those by ordinary microforming without ultrasonic vibration. The top and bottom views of star pyramids formed with and without ultrasonic vibration are shown in Figures 9 and 10, respectively. It is seen that not only the convex parts but also the concave parts are formed well, much better than expected. According to the top views, there does not seem to be a significant difference in the shape accuracy between the two pyramids formed with and without ultrasonic vibration. However, it is confirmed from the bottom views and more clearly from the magnified center of the bottom views that the traces of the tool path near the center of pyramid on the bottom surface is a nearly regular pentagon for ultrasonic assisted microforming, whilst it is heavily distorted for the ordinary microforming. Results in Figures 8–10 prove that the ultrasonic vibration can improve the shape accuracy of the formed pyramids. Other pyramids formed in this study using aluminum foil are a pyramid-like sunflower of D = 1.72 mm and a stepwise pyramid of D = 1.41 mm shown in Figure 11. They were formed well using ultrasonic incremental microforming.

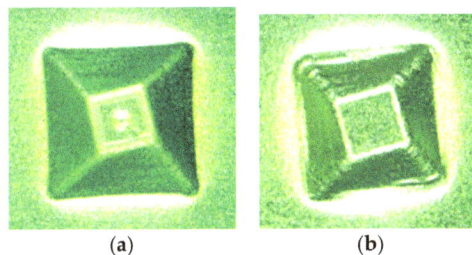

(a)   (b)

Figure 8. Top views of twisted pyramids formed: (**a**) with ultrasonic vibration; (**b**) without ultrasonic vibration.

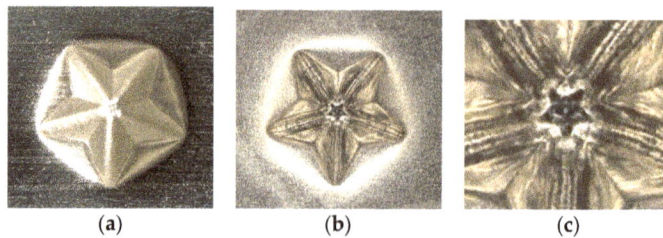

Figure 9. Top and bottom views of a star pyramid formed with ultrasonic vibration: (**a**) top view; (**b**) bottom view; (**c**) magnified center part of bottom view.

Figure 10. Top and bottom views of a star pyramid formed without ultrasonic vibration: (**a**) top view; (**b**) bottom view; (**c**) magnified center part of bottom view.

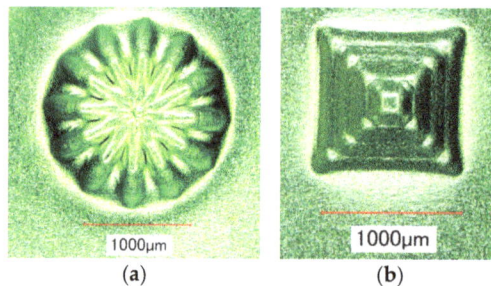

Figure 11. Other pyramids formed by ultrasonic incremental microforming: (**a**) pyramid like sunflower; (**b**) stepwise pyramid.

As described above, the ultrasonic vibration was effective in improving the shape accuracy of rather complicated pyramids such as a twisted pyramid and a star pyramid, whilst the ultrasonic vibration and laser heating were not so effective for truncated pyramids. In contrast, they were able to improve the shape accuracy of a truncated pyramid of 8 μm-thick stainless steel foil, which is much stiffer than 6.5 μm-thick aluminum foil. Two pyramids of stainless steel foil of $D = 1.0$ mm and $\theta = 45°$ were formed under NV and NH conditions and under V and H conditions. Their profiles measured with a confocal laser displacement meter along line AB of an attached figure are shown in Figure 12b,c. It is seen that the line of lateral face is connected with the base line via a curved line with a small radius under V and H conditions.

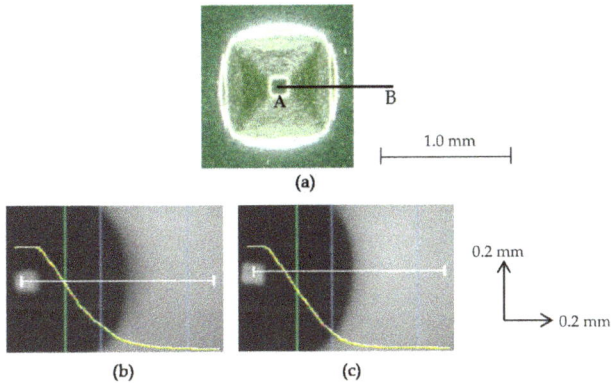

Figure 12. Profiles of pyramids of stainless steel foil under different conditions: (**a**) top view of a pyramid formed without ultrasonic vibration and heating; (**b**) surface profile along AB in the case of NV and NH; (**c**) surface profile along AB in the case of V and H.

The forming accuracy was evaluated based on distance δ from point C to the pyramid profile as shown in Figure 13. The value of δ was 43 μm and 26 μm for NV and NH conditions and V and H conditions, respectively. The theoretical value of the distance δ_{th} for perfect forming is given by

$$\delta_{th} = R\left[\sqrt{1 + \tan^2\left(\frac{\pi - 2\theta}{4}\right)} - 1\right] \tag{3}$$

and hence, the forming error can be defined by $\delta - \delta_{th}$. Because the value of δ_{th} is calculated to be 8.2 μm for $R = 100$ μm and $\theta = 45°$, the ultrasonic vibration and laser heating reduced the forming error by half. However, neither the ultrasonic vibration nor the laser heating increased the forming limit of a square pyramid of stainless steel foil: ε_c was 0.38 for (NV & NH) and (V & NH); it was slightly reduced to 0.35 for (NV & H) and (V & H). This is probably because it is difficult to heat the forming area uniformly due to very low heat conductivity of stainless steel.

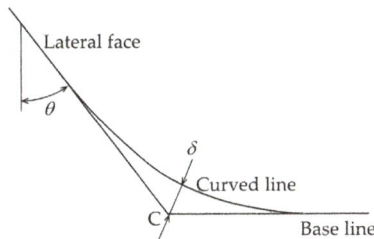

Figure 13. The forming accuracy based on distance δ from point C to the pyramid profile.

Thin shell pyramids of titanium foil of $D = 1.41$ mm and $\theta = 50°$ formed with and without ultrasonic vibration are shown in Figures 14 and 15, respectively. It was seen that wrinkles were caused on the lateral faces by incremental microforming when ultrasonic vibration was not applied. In contrast, they almost disappeared and feed marks of a single point tool were seen only on the back surface of a pyramid when ultrasonic vibration was applied. It should be noted that the ultrasonic vibration effectively removed the wrinkles in a difficult forming process of the twisted pyramid in Figure 8 as described above.

A truncated micropyramid of D = 283 µm formed with ultrasonic vibration is shown in Figure 16. Although the tool rotational speed ω was increased and table speed v_t and axial feed Δz were decreased for forming a smaller pyramid, its shape accuracy was not good, and partially distorted. It is seen that the radius of lateral edges was much larger than that of the forming tool and the feed marks of the tool on the back surface of the pyramid were disturbed. The size of a pyramid was reduced by a factor of five, but the foil thickness was not reduced. Hence, the bending stiffness relatively increased. This is a main reason that it was difficult to form a smaller micropyramid of titanium.

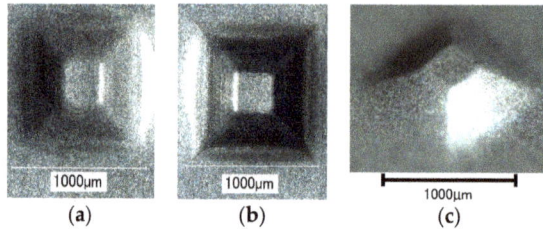

Figure 14. Titanium pyramids of D = 1.41 mm formed with ultrasonic vibration: (**a**) top view; (**b**) bottom view; (**c**) high angle view.

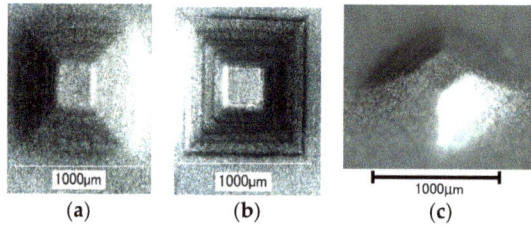

Figure 15. Titanium pyramids of D = 1.41 mm formed without ultrasonic vibration: (**a**) top view; (**b**) bottom view; (**c**) high angle view.

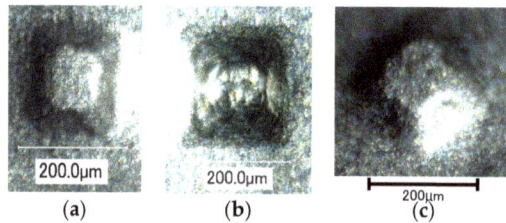

Figure 16. Titanium pyramids of D = 283 µm formed with ultrasonic vibration: (**a**) top view; (**b**) bottom view; (**c**) high angle view.

4. Conclusions

Single point incremental microforming of thin shell pyramids of aluminum, stainless steel, and titanium foils was conducted under the conditions of with and without ultrasonic vibration and with and without laser heating. It was found that the laser heating can improve the forming limit of a square pyramid of 6.5 µm-thick aluminum foil and it was more effective when ultrasonic vibration was applied in addition to laser heating. The elongation obtained under the assistance of ultrasonic vibration and laser heating was 73.4%, more than 10 times as large as that obtained by tensile test. It is confirmed that the assistance of ultrasonic vibration can improve the accuracy of rather complicated shape of pyramids such as a twisted pyramid and a star pyramid. It is also found that ultrasonic vibration improves the forming accuracy of stiff materials or difficult-to-work materials even if the shapes of micropyramids are simple.

Acknowledgments: The authors would like to extend their thanks to cutting tool manufacturers, OSG Corporation and NS Tool Co., Ltd. for the fabrication of microforming tools used in this study. This study is partly based on projects supported by Amada Foundation (AF-2014023) and JSPS Grant-in-Aids for Challenging Exploratory Research (24656097, 2012), Japan. These contributions are gratefully acknowledged.

Author Contributions: T.O. conceived and designed the experiments; M.H. completed the experimental apparatus, its controlling software, and forming experiments.

Conflicts of Interest: The authors declare no conflict of interest.

References

1. Dornfeld, D.; Min, S.; Takeuchi, Y. Recent advances in mechanical micromachining. *CIRP Ann.* **2006**, *55*, 745–768. [CrossRef]
2. Fissi, L.E.L.; Xhurdebise, V.; Francis, L.A. Effects of laser operating parameters on piezoelectric substrates micromachining with picosecond laser. *Micromachines* **2015**, *6*, 19–31. [CrossRef]
3. Zhang, J.; Li, Q.; Zhang, H.; Sui, Y.; Yang, H. Investigation of micro square structure fabrication by applying textured cutting tool in WEDM. *Micromachines* **2015**, *6*, 1427–1434. [CrossRef]
4. Raju, L.; Hiremath, S.S. A State-of-the-art review on micro electro-discharge machining. *Procedia Technol.* **2016**, *25*, 1281–1288. [CrossRef]
5. Wang, C.; Guo, B.; Shan, D. Friction related size-effect in microforming—A review. *Manuf. Rev.* **2014**, *1*, 23. [CrossRef]
6. Malinauskas, M.; Rekštytė, S.; Lukoševičius, L.; Butkus, S.; Balčiunas, E.; Pečiukaitytė, M.; Baltriukienė, D.; Bukelskienė, V.; Butkevičius, A.; Kucevičius, P.; et al. 3D microporous scaffolds manufactured via combination of fused filament fabrication and direct laser writing ablation. *Micromachines* **2014**, *5*, 839–858. [CrossRef]
7. Jeswiet, J.; Micari, F.; Hirt, G.; Bramley, A.; Duflou, J.; Allwood, J. Asymmetric single point incremental forming of sheet metal. *Ann. CIRP* **2005**, *54*, 88–114. [CrossRef]
8. Matsubara, S. A computer numerically controlled dieless incremental forming of a sheet metal. *J. Eng. Manuf.* **2001**, *215*, 959–966. [CrossRef]
9. Iseki, H. An approximate deformation analysis and FEM analysis for the incremental bulging of sheet metal using a spherical roller. *J. Mater. Process. Technol.* **2001**, *111*, 150–154. [CrossRef]
10. Obikawa, T.; Satou, S.; Hakutani, T. Dieless incremental micro forming of miniature shell objects of aluminum foils. *Int. J. Mach. Tools Manuf.* **2009**, *49*, 906–915. [CrossRef]
11. Sekine, T.; Obikawa, T. Single point micro incremental forming of miniature shell structures. *J. Adv. Mech. Des. Syst. Manuf.* **2010**, *4*, 543–557. [CrossRef]
12. Obikawa, T.; Hakutani, T.; Sekine, T.; Numajiri, S.; Matsumura, T.; Yoshino, M. Single-point incremental micro-forming of thin shell products utilizing high formability. *J. Adv. Mech. Des. Syst. Manuf.* **2010**, *4*, 1145–1156. [CrossRef]
13. Obikawa, T.; Sekine, T. Fabrication of miniature shell structures of stainless steel foil and their forming limit in single point incremental microforming. *Int. J. Autom. Techol.* **2013**, *7*, 256–262.
14. Beltran, M.; Malhotra, R.; Nelson, A.J.; Bhattacharya, A.; Reddy, N.V.; Cao, J. Experimental study of failure modes and scaling effects in micro-incremental forming. *J. Micro Nano Manuf.* **2013**, *1*, 031005. [CrossRef]
15. UACJ Foil Corporation. Available online: http://ufo.uacj-group.com/en/products/foil.html (accessed on 28 April 2017).
16. Arai, M.; Ogata, T. Development of small fatigue testing machine for film materials. *Trans. Jpn. Soc. Mech. Eng.* **2002**, *68*, 801–806. (In Japanese) [CrossRef]

micromachines

MDPI

Article

Slurry Injection Schemes on the Extent of Slurry Mixing and Availability during Chemical Mechanical Planarization

Matthew Bahr [1,*], Yasa Sampurno [1,2], Ruochen Han [1] and Ara Philipossian [1,2]

[1] Department of Chemical and Environmental Engineering, University of Arizona, Tucson, AZ 85721, USA; yasayap@email.arizona.edu (Y.S.); hanr@email.arizona.edu (R.H.); ara@email.arizona.edu (A.P.)
[2] Araca, Inc., Tucson, AZ 85718, USA
* Correspondence: mnbahr@email.arizona.edu; Tel.: +1-(520)-870-4133

Academic Editors: Hans Nørgaard Hansen and Guido Tosello
Received: 30 April 2017; Accepted: 26 May 2017; Published: 29 May 2017

Abstract: In this study, slurry availability and the extent of the slurry mixing (i.e., among fresh slurry, spent slurry, and residual rinse-water) were varied via three different injection schemes. An ultraviolet enhanced fluorescence technique was employed to qualitatively indicate slurry availability and its flow on the pad during polishing. This study investigated standard pad center area slurry application and a slurry injection system (SIS) that covered only the outer half of the wafer track. Results indicated that the radial position of slurry injection and the alteration of fluid mechanics by the SIS played important roles in slurry mixing characteristics and availability atop the pad. Removal rates were found to decrease with slurry availability, while a higher degree of slurry mixing decreased the fraction of fresh slurry and consequently lowered the removal rate. By using a hybrid system (i.e., a combination of slurry injection via SIS and standard pad center slurry application), the polishing process benefited from higher slurry availability and higher fraction of fresh slurry than the conventional pad center slurry application and the shorter SIS, individually. This work underscores the importance of optimum slurry injection geometry and flow for obtaining a more cost-effective and environmentally benign chemical mechanical planarization process.

Keywords: slurry availability; slurry injection system; slurry injection position; chemical mechanical planarization; CMP; slurry utilization efficiency

1. Introduction

Chemical mechanical planarization (CMP) is an enabling step in integrated circuit (IC) manufacturing for achieving local and global surface planarity through combined chemical and mechanical means. CMP has been widely used in the semiconductor manufacturing industry since 1985 [1]. Previous planarization technologies such as thermal flow and spin-on glass have been shown to provide adequate local planarity, but CMP is the only technique that additionally provides global planarity across the wafer surface [2]. A lack of global surface planarity results in tremendous difficulties during the following steps of the IC manufacturing process, such as lower photolithography and etch yields, greater step height variation, greater line-width variation, and the amplification of previous layer defects [3]. These variations critically impact and reduce both chip performance and overall device yield [1]. In addition to planarity, CMP must address issues concerning the surface quality of the wafer. Many academic and industry studies have shown that by controlling the process parameters and consumable sets, average surface roughness values ranging from 0.5 nm to 2 nm can be achieved [3–6].

The CMP process requires several consumables, such as polishing pad, pad conditioner disc, retaining ring, and slurry. The CMP slurry contains chemicals and nano-particles to support the chemical and mechanical actions of the removal process. In most CMP processes, slurry cost is a major component (up to 50%) of the overall cost of ownership [7,8]. Furthermore, slurry consumption can have serious environmental ramifications, as spent slurries contain hazardous chemicals as well as significant amounts of abrasive nano-particles. Even though the recovery and reuse of spent slurries has been investigated and adopted by a handful of IC manufacturers [9,10], we believe that the best solution is to optimize slurry usage by either reducing its flow rate to attain the same CMP performance (i.e., removal rate and defects) or by applying the same flow rate to achieve better performance as compared to the current process-of-record.

In most commercially available polishers, slurry is applied near the center of the pad, as shown in Figure 1a. As the pad continuously rotates, a large amount of fresh slurry flows radially off the pad surface without ever entering the pad–wafer interface. Such a flow pattern results in extremely low slurry utilization efficiencies [8,11]. Several other methods have been proposed for applying or injecting slurry onto the pad surface. For example, Mok has proposed an apparatus for spraying slurry directly on to the pad surface rather than dispensing the slurry via the standard pad-center stream application [12]. Chamberlin et al. have proposed a similar slurry injection technique involving spraying pressurized slurry over the pad through multiple nozzles [13]. Chiou et al. have also proposed a modified slurry dispensing apparatus employing "a plurality of adjustable nozzles" [14]. Finally, Chang has proposed a method for dispensing slurry through multiple nozzles above the pad near the pad–wafer interface which tends to promote slurry coverage over the entire wafer track [15]. It must be noted that while these methods help deliver fresh slurry to the pad, none of them prevent the mixing of spent slurry and residual rinse water with fresh slurry during polishing.

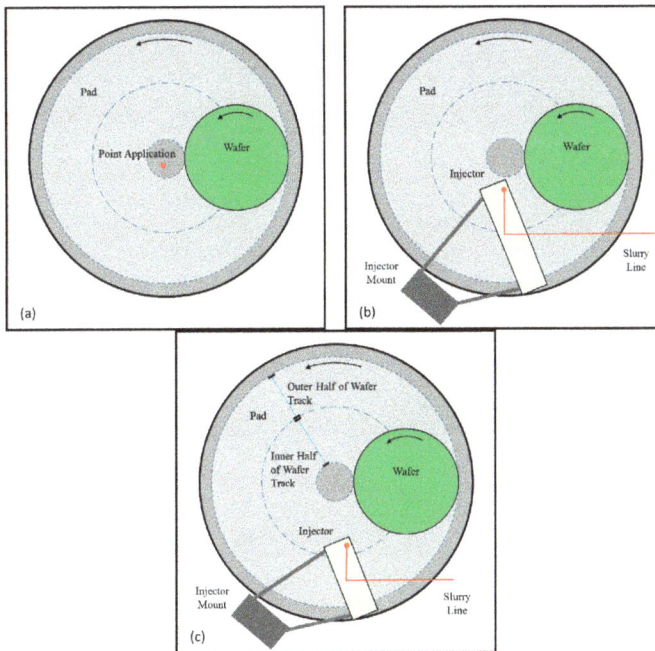

Figure 1. Top views of a polisher with (**a**) the standard slurry application method, (**b**) a slurry injection system (SIS) design that covers the whole wafer track, and (**c**) a SIS design that covers only the outside half of the wafer track.

Polishing processes continuously generate spent slurry on the pad surface. Spent slurries often contain pad debris (from the conditioning process and also from normal pad wear), diamond chips (that may get dislodged from diamond discs), and chemical by-products. Studies have shown that these contaminants can decrease material removal rate and increase wafer-level defects [11,16]. To mitigate such issues, large amounts of ultrapure water (UPW) are used to rinse the pad between polishes. Following pad rinsing, appreciable amounts of residual rinse water still reside on the land areas of the pad as well as inside the grooves. When fresh slurry is injected onto the pad during polishing, it mixes with the residual rinse water and gets diluted. As most industrially-relevant slurries result in lower removal rates when further diluted with water, over-the-pad mixing of water and slurry should be avoided. As such, it is fair to say that the current standard pad center slurry application method as well as the aforementioned slurry application or injection methods do not provide efficient slurry utilization. This has provided us with the opportunity to improve polishing performance.

As an alternate method, Meled et al. and Mu et al. investigated a slurry injection system (SIS) which is aimed to shorten the slurry mean residence time (MRT) on the pad surface [11,17]. In both studies, the SIS was placed adjacent to the wafer on the pad surface covering the whole wafer track, as shown in Figure 1b. The SIS facilitated the delivery of fresh slurry to the pad–wafer interface. In addition, the SIS effectively blocked the spent slurry and residual rinse water from re-entering the pad–wafer interface and therefore allowed a higher fraction of the fresh slurry to be delivered to the polishing region. As a result, the SIS achieved a significantly lower slurry mean residence time, higher removal rate, and lower polishing defects than the standard pad center area slurry application method [11,17].

It must be emphasized that Meled et al. and Mu et al. employed an SIS design that covered the whole wafer track on the surface of the polishing pad as shown in Figure 1b. As a matter of a fact, some CMP systems designed for high volume manufacturing have particular space restrictions that will not allow the implementation of SIS units that cover the entire wafer track on the surface of the pad. Such space restrictions are typically associated with oscillations of the conditioner and the wafer carrier head. As a continuation of both Meled and Mu's works, this study investigates an SIS design that covers only the outer half of the wafer track, as shown in Figure 1c. This study aims to understand if such an SIS unit—which may need to be shorter in length only because of possible space limitations on the pad—can provide similar benefits as the full-size SIS design. In addition, an ultraviolet enhanced fluorescence (UVEF) technique is employed to qualitatively measure the slurry availability on top of the pad prior to its entry into the pad–wafer interface. When compared to the standard pad center slurry application method, results help to confirm the main mechanism responsible for the enhanced removal rate associated with the SIS.

2. Materials and Methods

The standard pad center area slurry application method and a novel method for slurry injection system were used to apply slurry onto the pad surface. Figure 1a,c show the top views of a polisher with the standard slurry application method and the SIS design that only covers the outer half of the wafer track. For the standard pad center area slurry application method (what we will henceforth refer to as "Point Application" or "PA"), the slurry is applied above the pad center. For the SIS, the system consists mainly of an injector and an injector mount. The injector has a rectangular shape which is attached to the injector mount with the connecting rods. The bottom of the injector is in contact with the surface of the polishing pad. The mount is used to securely attach the entire SIS assembly to the polisher's frame. A single slurry inlet port is placed on top of the body which matches an outlet at its bottom y (at the trailing edge). Fresh slurry is introduced through the slurry feed line from the slurry tank where it flows into the inlet and then flows out into a channel machined into the bottom of the injector body. This channel helps to evenly spread the fresh slurry onto pad surface during polishing. A full description of the SIS can be found elsewhere [18]. In one test configuration, a hybrid slurry injection method is employed. The rationale behind this hybrid injection method is discussed in detail in Section 3 of this paper.

All wafer polishing was done on an Araca APD-800X polisher. Detailed description of this polisher may be found elsewhere [19]. A 3M A165 diamond disc was used to perform in-situ conditioning on an IC-1000 pad (manufactured by Dow) with a "K-groove" pattern. The conditioning down-force was set to 44.5 N. Each wafer was polished for 1 min at 27.6 kPa and 1.5 m/s. Before polishing, pad break-in was performed with the diamond disc for 60 min with DI water. The conditioning disc's rotation rate was set to 95 RPM, and its sweep frequency across the radius of the pad to 10 times per min. The diamond disc, pad, and wafer rotations were all counter clockwise. Pad break-in was then followed by pad seasoning, during which the shear force was monitored in real-time to ensure that stable values were achieved prior to any polishing with monitor wafers. It is important to note that even though we selected a relatively hard pad for the polishing tests, the application and utility of the SIS is by no means limited to hard pads.

The Semi-Sperse® 25 slurry (manufactured by Cabot Microelectronics, Aurora, IL, USA)—diluted with water to a final solids content of 12.5% by weight—was used as the polishing slurry. Slurry flow rates were set at 150 and 250 mL/min. Blanket silicon dioxide wafers (300-mm) were polished for all injection schemes. Before and after polishing each wafer, a reflectometer from SENTECH Instruments GmbH (Berlin, Germany) was used to measure the thickness of the silicon dioxide film, which allowed us to compute the average removal rate for each test. Within-wafer removal rate non-uniformity (WIWRRNU) as well as wafer-level large particle counts were not determined in this study. However, in an earlier work by our team [20] using similar consumables and process conditions, SIS yielded equivalent values for WIWRRNU ($3.9 \pm 0.6\%$ vs. $4.0 \pm 0.5\%$ for PA, 1-σ with 5 mm edge exclusion) and significantly lower wafer-level, greater than 0.5 micron, particle counts (174 ± 57 vs. 438 ± 155 for PA, 1-σ with 5 mm edge exclusion). Furthermore, it should be noted that our objective here was to polish silicon dioxide films, but the methods and techniques employed in our work are also applicable to other insulating (i.e., silicon nitride) as well as conducting (i.e., tungsten, tantalum, and copper) films.

In addition to the above series of polishing tests, an ultraviolet enhanced fluorescence (UVEF) technique was employed to qualitatively visualize slurry flow patterns and measure the availability of the slurry (i.e., its thickness) on the pad surface [21,22]. Before taking images, an embossed Politex pad (manufactured by Dow Electronic Materials, Newark, DE, USA)—which is softer compared to IC-1000—was conditioned using a 3M PB32A brush for 30 min with UPW at a conditioning force of 13.3 N. The Politex pad was employed because of its black color, not its mechanical properties, as in UVEF tests attaining superior color contrast between the fluorescing dye and the pad is critical. In all experiments, the pad was conditioned in-situ. The slurry for the UVEF experiments consisted of 1 volume part of Fujimi PL-7103 slurry, 4 volume parts of DI water, and 0.5 g/L of 4-methyl-umbelliferone. Pad rinsing was performed in between polishes using UPW. The experimental setup is shown in Figure 2. Ultraviolet (UV) light from two light-emitting diodes was projected onto the leading edge of the carrier head. As the slurry was tagged with a fluorescent dye (i.e., 4-methyl-umbelliferone), the UV light excited the dye in the slurry, causing it to fluoresce. The intensity of the emitted fluorescence was proportionate to the amount of the slurry (i.e., the thickness of the slurry film) [21,22]. A high-resolution charged coupled device camera was employed to record the emission of fluorescent light on the leading edge of the wafer carrier head. The images were then analyzed via a customized software written in LabVIEW [21,22].

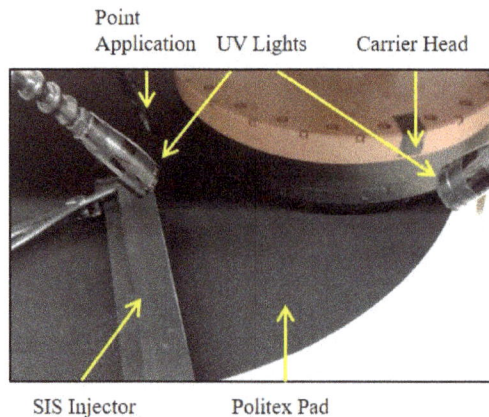

Figure 2. Experimental setup for ultraviolet enhanced fluorescence (UVEF) on an Araca APD-800 Polisher.

3. Results and Discussion

Several published reports indicated that 300-mm CMP processes typically employ slurry flow rates ranging from 250 mL/min to 300 mL/min to achieve optimum material removal rates (RR) [11,23–25]. In a separate study, Wang et al. showed that RR increases with the PA slurry flow rate, but it eventually reaches an asymptote where further increases in slurry flow rate no longer affect removal rate [26]. Philipossian et al. showed that the slurry utilization efficiency (defined as the portion of slurry that flows through the pad–wafer interface, divided by the total amount of slurry applied) actually decreases with an increase in slurry flow rate [8]. Simply increasing slurry flow rate until a certain level may help achieve higher RR in several cases, but at a disproportionality higher consumable cost. Therefore, in this study, polishing using the PA method at a slurry flow rate of 250 mL/min (henceforth referred to as "PA-250") is considered as our baseline process. Figure 3 summarizes the removal rates for all slurry injection methods (to be elaborated in detail later on in this section). As expected, PA at reduced slurry flow rate (i.e., 150 mL/min, henceforth referred to as "PA-150") yields lower RR than PA-250. Several published results have reported a similar finding, whereby removal rates are shown to decrease with slurry flow rate [11,26]. One reason for this observation is that at the reduced flow rate, less slurry is available to be transported to the pad–wafer interface [21,27]. Another reason is the degree of slurry mixing: as discussed in Section 1, during polishing, the fluid residing on the pad surface contains spent slurry, polishing by-products (i.e., pad debris and chemical by-products), and residual rinse water from the pad rinsing procedure performed between polishes which do not contribute to RR [11,27]. The fresh slurry injected on top of the pad surface gets diluted by mixing with spent slurry and residual rinse water. The net effect is the reduction of the fraction of fresh slurry delivered to the polishing region (i.e., between the pad and the wafer) where the removal mechanism occurs. At reduced slurry flow rate (i.e., the PA-150), the fraction of fresh slurry becomes even lower, thus leading to a lower RR [11,17,27]. As a matter of fact, the traditional PA method does not provide a mechanism to promote a higher fraction of fresh slurry. Based on the explanation above, we can infer that both the degree of slurry mixing and slurry availability affect RR during polishing processes.

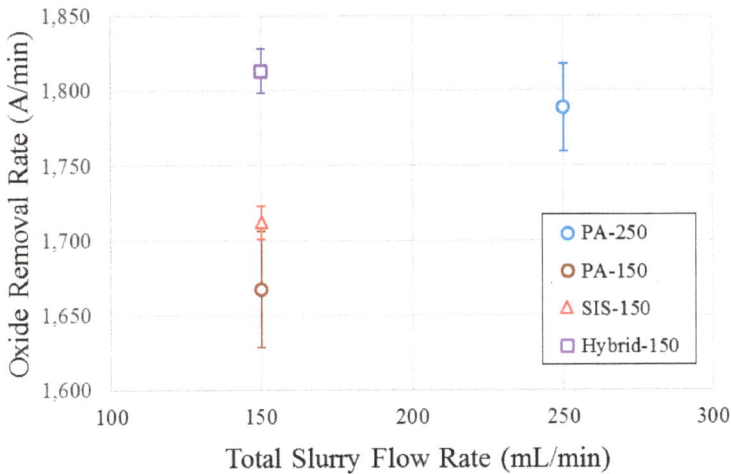

Figure 3. Summary of removal rate data. PA-250: polishing using the PA method at a slurry flow rate of 250 mL/min; PA-150: polishing using the PA method at a slurry flow rate of 150 mL/min; SIS: polishing using the SIS method at a slurry flow rate of 150 mL/min; Hybrid-150: polishing using the hybrid method (SIS at a slurry flow rate of 50 mL/min + PA at a slurry flow rate of 100 mL/min) at a total slurry flow rate of 150 mL/min.

Previous studies showed that slurry mean residence time (MRT) of a certain CMP process is an indicator for the degree of slurry mixing, such that a higher value of MRT means more mixing between the freshly injected slurry and spent slurry, as well as residual rinse water [11,17]. Therefore, lower values of MRT are desirable in CMP because more fresh slurry is being delivered to the pad–wafer interface at a faster rate. Meled et al. and Mu et al. have shown that SIS significantly decreases slurry MRT compared to the PA method [11,17]. Furthermore, Mu et al. concluded that the dispersion number is lower with SIS, which accounts for the lower MRT [17]. The mechanism can be further explained with the UVEF images shown in Figure 4. With the PA injection scheme, a thick bow wave containing spent slurry and residual rinse water is formed directly at the leading edge of the wafer carrier head, as shown in Figure 4a. Consequently, the PA method allows more spent slurry and residual rinse water to re-enter the wafer–pad interface. In contrast, as the SIS is placed in front of the leading edge of the wafer carrier head, it prevents spent slurry and residual rinse water from re-entering the pad–wafer interface during polishing. Figure 4b shows the UVEF image of a polishing process using an SIS design that covers only the outer half of the wafer track. Similar to a regular SIS design (i.e., covering the whole wafer track), the bow wave was formed at the leading edge of the SIS during polishing. Due to the centrifugal force of the platen and wafer rotation, most of the polishing by-products, spent slurry, and residual rinse water dominantly reside closer to the edge of the pad rather than to the center of the pad. Therefore, having a smaller SIS design that covers only the outer half of the pad is still desirable. As shown in Figure 4b, an SIS design that covers the outer half of the wafer track can effectively block polishing by-products, spent slurry, and residual rinse water from re-entering the pad–wafer interface. The thick bow wave formed at the leading edge of the SIS closer to the edge of the pad confirms that the spent slurry and residual rinse water are effectively blocked from re-entering the pad–wafer interface and guided off from the pad surface to the drainage. Furthermore, by using the SIS, the fresh slurry is less diluted with the spent slurry and residual rinse water, leading to a higher fraction of fresh slurry that enters the pad–wafer interface.

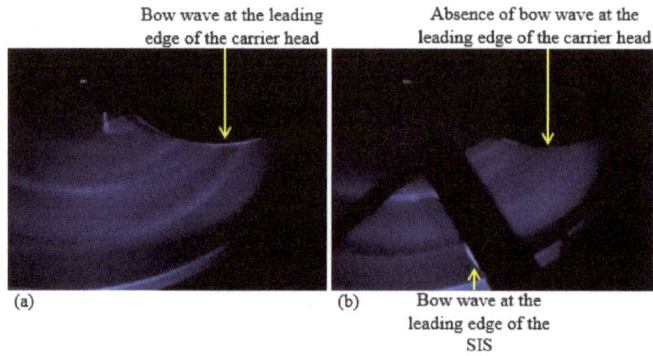

Figure 4. Bow wave formation on the polishing pad using (**a**) Point Application (PA) method and (**b**) the SIS.

SIS increases the slurry availability and reduces the degree of mixing of the fresh slurry with the spent slurry and residual rinse water. As a result, using the SIS at a reduced slurry flow rate can achieve similar removal rates compared to PA [11,17,21]. Previous studies on the SIS (that fully cover the whole wafer track) showed that SIS at a slurry flow rate of 150 mL/min achieved the same removal rate as PA at a slurry flow rate of 250 mL/min [11,17,21]. Referring back to Figure 3, using an SIS that only covers the outer half of the wafer track at a flow rate of 150 mL/min (referred to as "SIS-150") yields an RR that is significantly lower than the PA-250 and only slightly higher than the PA-150. In this configuration, the slurry injection port on the SIS coincides with the center of the wafer track. Due to rotation and centrifugal forces of both platen and wafer, the injected fresh slurry is then mainly distributed on the outer half of the wafer track, as illustrated by red color in Figure 5. Therefore, injecting the slurry in the center of the wafer track reduces slurry coverage in the pad–wafer interface, as entering fresh slurry is now reduced to one-half of what was observed when PA and full SIS designs were used. Furthermore, centrifugal forces act more rapidly to pull the slurry off of the pad surface as the fresh slurry in injected closer to the edge of the pad. Therefore, in this configuration, while SIS is still effective in blocking the spent slurry and residual rinse water, slurry availability is greatly reduced because it only covers the outer half of the wafer track. To increase slurry availability on the inner half of the wafer track while still taking the benefit of SIS to block the spent slurry and residual rinse water to re-enter the pad–wafer interface, a hybrid system is proposed as shown in Figure 6.

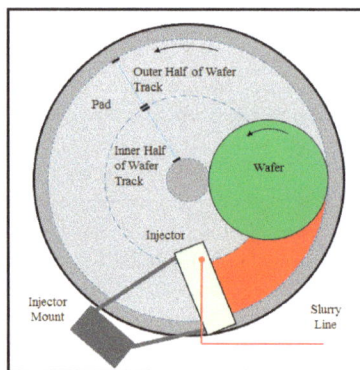

Figure 5. Coverage of fresh slurry on wafer track (red) using slurry injection system with slurry injection point coinciding with the center of the wafer track.

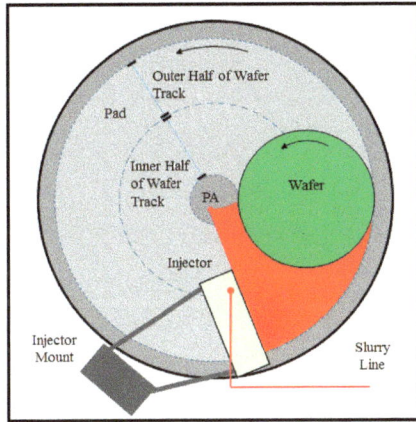

Figure 6. Coverage of fresh slurry on wafer track (red) using a hybrid system.

Figure 6 illustrates the hybrid system that combines both SIS and PA (what we will henceforth refer to as "Hybrid"). In this case, fresh slurry streams are injected contemporaneously to the SIS as well as to point application (PA) at 50 and 100 mL/min, respectively. The total slurry flow rate remains the same at 150 mL/min (i.e., 40 percent reduction from the 250 mL/min using the regular PA method). The main purpose for having slurry injected through the PA is to facilitate the availability of fresh slurry in the inner half of the wafer track. In the inner half of the wafer track, slurry dilution (with spent slurry and residual rinse water) is unavoidable, since SIS does not cover this region. However, the dilution is expected to be significantly less pronounced than the outer half of the wafer track. Similarly, the location of the SIS is kept constant (i.e., covering the outer half of the wafer track) in order to effectively block the spent slurry and residual rinse water from re-entering the pad–wafer interface as previously explained. At the same time, slurry is injected via SIS to ensure enough availability of fresh slurry on the outer half of the wafer track.

In summary, the hybrid system is expected to provide enough availability of fresh slurry covering the whole wafer track and to still cause the removal of spent slurry and residual rinse water. As shown in Figure 3, the Hybrid-150 yields an RR that is comparable to or slightly higher than the PA-250. Compared to PA, the hybrid system accommodates the delivery of fresh slurry, as it covers the whole wafer track and increases the fraction of fresh slurry delivered to the pad–wafer interface (i.e., polishing region where the removal mechanism occurs) by squeegeeing (i.e., wiping) off most of the spent slurry and residual rinse water.

To further explain the above results, a UVEF technique was employed to qualitatively measure slurry availability on top of the pad surface. Table 1 summarizes the average UVEF intensity throughout the polishing time, obtained by analyzing UV-enhanced fluid film on the two regions depicted in Figure 7. Regions 1 and 2 are located on the inner and outer half of the wafer track, respectively. Both regions are located on the pad prior to the entering of the slurry in the pad–wafer interface. With the SIS installed, both regions are intentionally set between the SIS and wafer carrier head. For a fair comparison among PA, SIS, and hybrid systems, the location for both Regions 1 and 2 was kept the same. It is important to note that UVEF intensity in this study does not measure an exact fluid film volume, but rather a relative slurry availability based on findings in a previous study [21]. The UVEF intensity increases with slurry availability and vice versa. In addition, such a technique is not intended to exactly quantify the composition of the fluid film (i.e., spent slurry, fresh slurry, and residual rinse water). As shown in Table 1, PA-250 results in the highest UVEF intensity measured in both Regions 1 and 2. This is intuitive because this particular slurry application method uses the highest flow rate and has no mechanism for squeegeeing off the spent slurry.

Table 1. Summary of ultraviolet enhanced fluorescence (UVEF) images analysis.

Region 1		
Slurry Injection Scheme	**Average UVEF Intensity (A.U.)**	**Standard Deviation (A.U.)**
PA-250	53.5	1.9
PA-150	48.8	3.3
SIS-150	27.0	1.2
Hybrid-150	44.7	2.5
Region 2		
Slurry Injection Scheme	**Average UVEF Intensity (A.U.)**	**Standard Deviation (A.U.)**
PA-250	66.0	3.5
PA-150	57.1	3.4
SIS-150	41.8	2.4
Hybrid-150	59.9	2.2

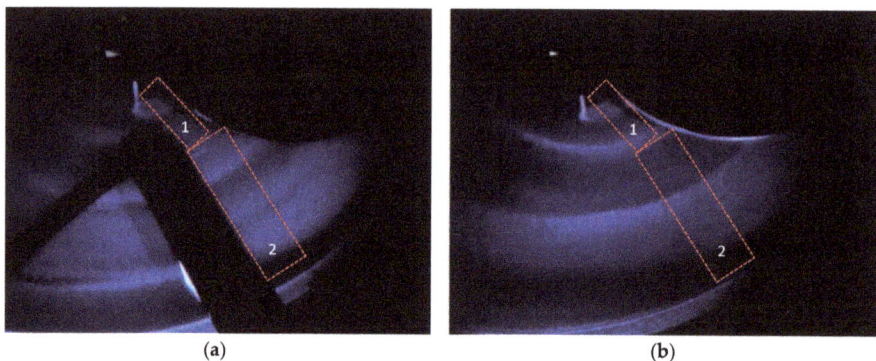

(a)   (b)

Figure 7. Regions 1 and 2 for UVEF analysis on the polishing pad: (**a**) with the SIS, and (**b**) without the SIS.

Compared to PA-250, PA-150 decreases the UVEF image intensity by approximately 9% and 14% in Regions 1 and 2, respectively. Such decreases are expected due to reductions in the injection of fresh slurry. It must be noted that a controlled pad rinsing procedure with UPW (that contains no fluorescence dye) is performed prior to every polish. As such, the amount of residual rinse water on the pad is the same prior to every polish on both the PA-250 and PA-150. At a reduced slurry flow rate (i.e., PA-150), the freshly injected slurry is diluted more with the residual rinse water. Furthermore, a lower slurry flow rate takes a longer time to replace the residual rinse water with the dyed slurry. Since the mechanisms of slurry injection and fluid removal are essentially the same between PA-250 and PA-150, this reduction in image intensity associated with PA-150 confirms a lower fraction of fresh slurry to overall fluid on the pad during polishing compared to PA-250. As a result, the RR of PA-150 is lower than that of PA-250.

Figure 3 shows that the RR of the SIS-150 is marginally higher than PA-150. In the meantime, Table 1 shows that when changing the slurry application mechanism from PA-150 to SIS-150, the UVEF image intensity decreases significantly by approximately 45% and 27% in Regions 1 and 2, respectively. Compared to PA, SIS incorporates a mechanism that squeegees and wipes off the residual rinse water as well as the spent slurry that contains the dye. During polishing, the spent slurry has a reduced polishing capability (in terms of RR) compared to fresh slurry, but the dye component itself is not degraded in terms of fluorescence intensity. By effectively squeegeeing off the spent slurry, SIS-150 is artificially showing less slurry availability compared to PA-150 as shown by its lower UVEF image intensity. In fact, SIS prevents the spent slurry from re-entering the pad–wafer region, and therefore the UVEF image intensity associated with SIS in Regions 1 and 2 can be regarded as a much higher fraction of the fresh slurry (i.e., less dilution with spent slurry). Compared to PA-150, the significant drop of UVEF intensity with SIS-150 in Region 1 is attributed to the absence of slurry injection in the inner half of the wafer track. As a result, only spent slurry contributes to the UVEF intensity with the SIS-150 in Region 1. Region 2 of the SIS-150 has a significantly higher intensity than Region 1 because the fresh slurry is injected through the SIS that covers the outer half of the wafer track (i.e., Region 2). The UVEF technique shows that even though SIS-150 can effectively wipe off the spent slurry and residual rinse water, the fresh slurry is mainly available only to the outer half of the wafer track.

Compared to SIS-150, the Hybrid-150 increases the UVEF intensity of Regions 1 and 2 by approximately 40% and 30%, respectively. In Region 1, the sharp increase in intensity is attributed to the addition of slurry injection on the inner half of the wafer track. As a result, it increases the fraction of fresh slurry to the overall fluid and hence the slurry availability in the inner half of the wafer track. The increase of intensity in Region 2 also outlines the effect of full wafer track slurry coverage, as illustrated in Figure 6. The rotation of the platen–wafer ejects the slurry from the inner half of the wafer track toward the outer half of the wafer track. Such a mechanism, combined with the fresh slurry injected through the injector, increases slurry availability in Region 2 and thereby causes higher UVEF intensity. Using the hybrid method, the combined mechanisms of the full slurry coverage and the squeegee effects (of spent slurry and residual rinse water) increase the fraction of fresh slurry while maintaining slurry availability on top of the polishing pad during polishing.

4. Conclusions

This work has shown that slurry availability and the extent of slurry mixing (i.e., among fresh slurry, spent slurry, and residual rinse water) dramatically influence removal rates. The ultraviolet enhanced fluorescence (UVEF) technique showed that injecting fresh slurry solely on the center of wafer track reduces slurry availability in the pad–wafer interface, as it only covered the outer half of the wafer track. Removal rates were found to decrease with slurry availability accordingly. A higher degree of slurry mixing decreased the fraction of fresh slurry, and consequently lowered the removal rate. In this study, a novel slurry injection system was installed on top of the polishing pad that covered only the outer half of the wafer track. UVEF technique confirmed that most of the fluid (i.e., slurry and residual rinse water) predominantly resided closer to the edge of the pad rather than near the center of the pad. Therefore, the SIS that covered only the outer half of the wafer track was still effective in blocking the spent slurry and the residual rinse water. This mechanism facilitated the increase in the fraction of fresh slurry on the polishing pad during polishing. In contrast, standard pad center area slurry application did not have a similar mechanism to block the spent slurry and residual rinse water.

Results further indicated that slurry injection position and the novel SIS played important roles in slurry mixing characteristics and slurry availability on top of the pad. Injecting fresh slurry at a higher flow rate and at a location closer to the pad center area generally increased slurry availability. By using a hybrid system (i.e., a combination of slurry injection through the shorter SIS and a standard pad center area slurry application), the polishing process benefited from higher slurry availability and higher fraction of fresh slurry than the pad center area slurry application and the shorter SIS, individually. This was verified by the UVEF technique.

Acknowledgments: The authors would like to express their deepest gratitude to the University of Arizona's department of Chemical and Environmental Engineering for covering the publication costs of this article.

Author Contributions: A.P. and Y.S. conceived the experiments; M.B. and R.H. designed and subsequently performed the experiments; M.B. and Y.S. analyzed the data; A.P. contributed the slurry, wafers, and all polishing and analysis tools; All authors contributed suggestions and participated in meaningful discussions regarding the experiment and manuscript writing; M.B. wrote the manuscript; Y.S., R.H., and A.P. edited the manuscript.

Conflicts of Interest: The authors declare no conflict of interest.

References

1. Landis, H.; Burke, P.; Cote, W.; Hill, W.; Hoffman, C.; Kaanta, C.; Koburger, C.; Lange, W.; Leach, M.; Luce, S. Integration of Chemical-Mechanical Polishing into CMOS Integrated Circuit Manufacturing. *Thin Solid Films* **1992**, *220*, 1–7. [CrossRef]
2. Ali, I.; Roy, S.R.; Shinn, G. Chemical-Mechanical Polishing of Interlayer Dielectric: A Review. *Solid State Technol.* **1994**, *37*, 63.
3. Zantye, P.B.; Kumar, A.; Sikder, A.K. Chemical Mechanical Planarization for Microelectronics Applications. *Mater. Sci. Eng. R Rep.* **2004**, *45*, 89–220. [CrossRef]
4. Zhou, C.; Shan, L.; Hight, J.R.; Danyluk, S.; Ng, S.H.; Paszkowski, A.J. Influence of Collodial Abrasive Size on Material Removal Rateand Surface Finish in SiO_2 Chemical Mechanical Polishing. *Tribol. Trans.* **2002**, *45*, 232–238. [CrossRef]
5. Choi, G.; Lee, K.; Kim, N.; Park, J.; Seo, Y.; Lee, W. CMP Characteristics and Optical Property of ITO Thin Film by Using Silica Slurry with a Variety of Process Parameters. *Microelectron. Eng.* **2006**, *83*, 2213–2217. [CrossRef]
6. Basim, G.B.; Adler, J.J.; Mahajan, U.; Singh, R.K.; Moudgil, B.M. Effect of Particle Size of Chemical Mechanical Polishing Slurries for Enhanced Polishing with Minimal Defects. *J. Electrochem. Soc.* **2000**, *147*, 3523–3528. [CrossRef]
7. Holland, K.; Hurst, A.; Pinder, H. Improving Cost of Ownership and Performance of CMP Process and Consumables. *Micro* **2002**, *20*, 26–30.
8. Philipossian, A.; Mitchell, E. Slurry Utilization Efficiency Studies in Chemical Mechanical Planarization. *Jpn. J. Appl. Phys.* **2003**, *42*, 7259–7264. [CrossRef]
9. Kettler, D.S. *Reclaim, Recycle and Reuse of Oxide Slurry for the CMP Process*; SEMICON: Taipei, Taiwan, 2013.
10. Amoroso, N.; Tolla, B.; Boldridge, D. CMP Slurry Recycling System and Methods. U.S. Patent Application 20,120,042,575, 23 February 2012.
11. Meled, A.; Zhuang, Y.; Sampurno, Y.A.; Theng, S.; Jiao, Y.; Borucki, L.; Philipossian, A. Analysis of A Novel Slurry Injection System in Chemical Mechanical Planarization. *Jpn. J. Appl. Phys.* **2011**, *50*, 05EC01:1–05EC01:5. [CrossRef]
12. Mok, P. Apparatus for Dispensing Slurry. U.S. Patent 5,964,413, 12 October 1999.
13. Chamberlin, T.; Miller, M.; Walton, G. Slurry Injection Technique for Chemical Mechanical Planarization. U.S. Patent 5,997,392, 7 December 1999.
14. Chiou, W.; Chen, Y.; Shih, T.; Jang, S. Slurry Dispenser Having Multiple Adjustable Nozzles. U.S. Patent 6,398,627, 4 Jun 2002.
15. Chang, W. Methods for Enhancing Within-Wafer CMP Uniformity. U.S. Patent 6,929,533, 16 August 2005.
16. Kwon, T.; Cho, B.; Venkatesh, R.; Park, J. Correlation of Polishing Pad Property and Pad Debris on Scratch Formation During CMP. In Proceedings of the International Conference on Planarization/CMP Technology, Grenoble, France, 15–17 October 2012; pp. 391–396.
17. Mu, Y.; Han, R.; Sampurno, Y.; Zhang, Y.; Borucki, L.; Philipossian, A. Mean Residence Time and Dispersion Number Associated with Slurry Injection Methods in Chemical Mechanical Planarization. *ECS J. Solid State Sci. Technol.* **2016**, *5*, P155–P159. [CrossRef]
18. Borucki, L.; Sampurno, Y.; Philipossian, A. Method and Device for the Injection of CMP Slurry. U.S. Patent 8,845,395, 30 September 2014.
19. Fujikoshi Machinery Corporation. *APD–800 Polisher and Tribometer.* Available online: http://www.aracainc.com/media/pubs/Polisher_and_Tribometer_300mm.pdf (accessed on 4 February 2016).
20. Philipossian, A.; Borucki, L.; Sampurno, Y.; Zhuang, Y. Novel Slurry Injection System for Improved Slurry Flow and Reduced Defects in CMP. *Solid State Phenom.* **2014**, *219*, 143–147. [CrossRef]
21. Liao, X.; Sampurno, Y.; Zhuang, Y.; Philipossian, A. Effect of Slurry Application/Injection Schemes on Slurry Availability during Chemical Mechanical Planarization (CMP). *Electrochem. Solid State Lett.* **2012**, *15*, H118–H122. [CrossRef]
22. Liao, X.; Sampurno, Y.; Zhuang, Y.; Rice, A.; Sudargho, F.; Philipossian, A.; Wargo, C. Effect of Retaining Ring Slot Designs and Polishing Conditions on Slurry Flow Dynamics at Bow Wave. *Microelectron. Eng.* **2012**, *98*, 70–73. [CrossRef]

23. Jiao, Y.; Zhang, Y.; Wei, X.; Sampurno, Y.; Meled, A.; Theng, S.; Cheng, J.; Hooper, D.; Moinpour, M.; Philipossian, A. Pad Wear Analysis during Interlayer Dielectric Chemical Mechanical Planarization. *ECS J. Solid State Sci. Technol.* **2012**, *1*, N103–N105. [CrossRef]

24. Kangda, Y.; Shengli, W.; Yuling, L.; Chenwei, W.; Xiang, L. Evaluation of Planarization Capability of Copper Slurry in the CMP Process. *J. Semicond.* **2013**, *34*, 036002:1–036022:4. [CrossRef]

25. Jiao, Y.; Sampurno, Y.; Zhang, Y.; Wei, X.; Meled, A.; Philipossian, A. Tribological, Thermal, and Kinetic Characterization of 300-mm Copper Chemical Mechanical Planarization Process. *Jpn. J. Appl. Phys.* **2011**, *50*, 05EC02:1–05EC02:6. [CrossRef]

26. Wang, Y.G.; Zhang, L.C.; Biddut, A. Chemical Effect on the Material Removal Rate in the CMP of Silicon Wafers. *Wear* **2011**, *270*, 312–316. [CrossRef]

27. Sampurno, Y.; Borucki, L.; Philipossian, A. Effect of Slurry Injection Position on Slurry Mixing, Friction, Removal Rate, and Temperature in Copper CMP. *J. Electrochem. Soc.* **2005**, *152*, G841–G845. [CrossRef]

micromachines

MDPI

Article

Fabrication of Functional Plastic Parts Using Nanostructured Steel Mold Inserts

Nicolas Blondiaux [1,*], Raphaël Pugin [1], Gaëlle Andreatta [1], Lionel Tenchine [2], Stéphane Dessors [2], Pierre-François Chauvy [3], Matthieu Diserens [3] and Philippe Vuillermoz [4]

[1] Centre Suisse d'électronique et de Microtechnique CSEM, 1 rue Jaquet-Droz, CH-2002 Neuchatel, Switzerland; rpu@csem.ch (R.P.); gaa@csem.ch (G.A.)

[2] Centre Technique Industriel de la Plasturgie et des Composites-IPC, 2 rue Pierre et Marie Curie, 01100 Bellignat, France; Lionel.TENCHINE@ct-ipc.com (L.T.); Stephane.DESSORS@ct-ipc.com (S.D.)

[3] Micropat SA, 30 Côtes-de-Montbenon, CH-1003 Lausanne, Switzerland; pf@micropat.ch (P.-F.C.); mat@micropat.ch (M.D.)

[4] Vuillermoz SAS, 5 rue du Tomachon, 39200 Saint-Claude, France; pv-vuillermoz@orange.fr

* Correspondence: nbx@csem.ch; Tel.: +41-32-720-55-38

Academic Editors: Hans Nørgaard Hansen, Guido Tosello, Andreas Richter and Nam-Trung Nguyen
Received: 6 April 2017; Accepted: 25 May 2017; Published: 6 June 2017

Abstract: We report on the fabrication of sub-micro and nanostructured steel mold inserts for the replication of nanostructured immunoassay biochips. Planar and microstructured stainless steel inserts were textured at the sub-micron and nanoscale by combining nanosphere lithography and electrochemical etching. This allowed the fabrication of structures with lateral dimensions of hundreds of nanometers and aspect ratios of up to 1:2. Nanostructured plastic parts were produced by means of hot embossing and injection molding. Surface nanostructuring was used to control wettability and increase the sensitivity of an immunoassay.

Keywords: nanostructures; hot embossing; injection molding; polymer; 3D; steel; mold; wettability; immunoassay

---

1. Introduction

There is a growing trend for the fabrication of smart products with novel functionalities or enhanced performances. One route to achieve this goal is an accurate control of surface properties. Surface chemistry, topography and a combination of both can be engineered and optimized for specific applications. Surfaces with controlled topographies have, for example, been manufactured to reduce the adhesion of bacteria or living adherent cells [1,2] and to control friction and adhesion between surfaces [3]. The effect of surface roughness on wettability is a widely studied field [4]. Mimicking the well-known lotus effect has been the focus of many studies for the fabrication of superhydrophobic surfaces with controlled wetting states, for example, Cassie–Baxter vs. the Wenzel state [5,6]. As reported by Martines et al., the processing of silicon-based materials using advanced lithographic techniques allows the design and fabrication of surfaces with a high degree of control over surface chemistry and topography [7]. Plastic parts with controlled wettability have also been produced using replication techniques such as hot embossing and injection molding, however, the production of highly liquid-repellent plastic surfaces without any surface treatment remains challenging [8]. Another field of application relevant to this study is the use of surface micro- and nanostructuring to enhance the performance of sensors for biomedical and point-of-care diagnostics [9]. Highly sensitive plasmonic [10,11] and surface enhanced Raman spectroscopy (SERS) sensors have been produced by optimizing surface structures on metallic layers, with enhancement factors of up to 108 compared to on flat surfaces [12]. Another way to improve the sensitivity of sensors is to increase the specific area of the sensing element. This has been shown by Ingham et al., who used anodized alumina to create highly porous sensors with surface areas two to three orders of magnitude greater than flat surfaces [13]. Kim et al. also reported a significant increase in the fluorescence intensity (four times greater compared to a flat surface) after fabricating quartz nanopillars on the surface of a DNA biosensor [14]. Such an effect was also demonstrated by Kuwabara et al. [15] on polystyrene immunoassay chips using a nanoimprinting process. By using a specific elongation process during nanoimprinting, high aspect ratio pillars were produced, giving rise to a 34-fold increase in the fluorescence intensity.

The examples mentioned above confirm the potential benefits of surface micro- and nanostructuring and show that various techniques have been used to produce specific structures in different materials. One of the objectives of this study was to produce nanostructured, functional parts using replication techniques such as hot embossing and injection molding. One of the key points, therefore, is the fabrication of nanostructured mold inserts. Although the process chain developed for the production of CDs and DVDs is well established for the fabrication of nanostructured nickel stampers, there is an increasing interest to produce steel mold inserts. The main motivations are the ease of integration in standard injection molds, the wide range of materials with better durability and the possibility of combining different techniques to process inserts at different length scales without being limited by the thickness of the electroformed stamper. Conventional techniques such as micromilling, laser ablation and wire electrical discharge machining (EDM) are well established for the microstructuring of steels [16,17]. These are cost effective and allow the fabrication of high aspect ratio structures on freeform shapes, but have a resolution limited to few micrometers. Alternatively, advanced lithography techniques have been developed for the microelectronic industry during the last few decades. UV, X-ray, interference and e-beam lithography, when combined with other microfabrication processes such as thin film deposition and etching, have paved the way for the fabrication of nanostructured surfaces and devices [18,19]. Emerging bottom-up approaches such as block copolymer and nanosphere lithography have also appeared with the common goal of fabricating smaller structures [20,21]. These techniques are state of the art with regard to resolution, but they are limited to planar substrates and have mainly been applied to silicon-based materials. When combined with electroforming, structured nickel stamps form the basis of the production of CDs and DVDs. Several studies report the use of photolithography to fabricate structured steel surfaces. Compared to standard microstructuring techniques such as milling or EDM, this involves the fabrication of an etch mask, followed by the etching of the substrate. In photochemical etching, standard photolithography is combined with chemical etching for the fabrication of a micropart or for surface microstructuring as shown by Hao et al., Masuzawa et al. and Mason et al. [22–24]. However, one drawback of chemical etching is the significant increase in the surface roughness of the etched surface, which results from the chemical etching solution used and the microstructure of the steel. This can be overcome by using an electrochemical etching technique, which leads to a mirror-polished surface in the etched areas. Landolt et al. made comprehensive reviews on the effect of the main parameters affecting electrochemical etching for titanium and stainless steel surfaces [25–27]. Usually limited to an isotropic profile, electrochemical etching has also been used in a sequential Bosch-like process to create higher aspect ratio structures, as reported by Shimizu et al. [28]. Finally, physical etching techniques have also been used for the fabrication of nanostructured steel surfaces. An alternative to wet etching techniques has been proposed by Al-Azawil et al., who combined photolithography and ion beam etching for the surface structuring of injection mold inserts. The very low selectivity of ion beam etching allowed homogeneous etching of the steel surface despite the heterogeneity of the material (Cr, Ni content, presence of MnS inclusions, steel microstructure) [29]. This technique was also used by Kurhihara et al. with a layer of nanoparticles as an etch mask [30]. Although the nanoparticle layers used as etch masks are not as well defined as etch masks made by photolithography, this approach allowed the processing of curved injection molding dies for the fabrication of optical lenses with improved antireflectivity.

When specifically considering the surface structuring of injection mold inserts, the process chain developed for CDs and DVDs remains state of the art for the injection molding of nanostructured plastic parts. However, several limitations remain concerning the ease of integration of the nickel inserts in the molds, its durability, its compatibility with other techniques to make hierarchical structures, the processing of non-planar surfaces and the presence of nickel which is banned for medical applications. An alternative is the direct processing of steel grades being used for mold manufacturing. This allows standard techniques such as milling, EDM and laser ablation to be used to control the overall shape and macro-/microstructures of the insert, and surface micro- and nanostructuring to be added to the standard mold manufacturing chain. This approach also gives greater flexibility regarding the materials of the inserts and the means of integration into the existing mold. Therefore, the overall objective of this study is to develop new processes to engineer the surface roughness of steel inserts compatible with injection molding and to apply them to the production of nanostructured plastic parts. The main challenge is to apply surface structures with typical lateral dimensions (a few hundred nanometers) onto the multilevel microfeatures of a stainless steel insert. In achieving this goal, most of the techniques mentioned above would face limitations. Although they have sufficiently high resolution, standard techniques such as photo-, e-beam or interference lithography would not be feasible due to the tridimensional shape of the part to treat. Standard microstructuring techniques, such as EDM or micromilling, are 3D compatible but do not have a sufficiently high resolution. The process flow used for the fabrication of CDs and DVDs would be one way to produce multilevel structures but a major objective of this study is to propose an alternative to nickel inserts for injection molding for the aforementioned reason.

In this study, a new process chain for the surface structuring of steel inserts has been developed. The technique, based on the combination of nanosphere lithography and electrochemical etching, has been used to fabricate structured stainless steel inserts with sub-micro-/nanofeatures. The choice of these techniques was mainly due to their compatibility with 3D parts. The final goal of the project was the production of a bio-diagnostic platform capable of performing immunoassays with increased sensitivity. The functional part was a microscope slide with an array of detection spots (micropillars) located at the bottom of a microchannel. The objective was to introduce micro-/nanostructures on top of the detection spots to control the functionalization of the spots during the immunoassay (via control of wetting) and to enhance the fluorescence signal. Nanostructured 2D surfaces were produced by hot embossing as references, and nanostructured bio-diagnostic platforms were produced by injection molding. The effect of surface structuring on wettability was characterized by means of water contact angle measurements and a model immunoassay was carried out to investigate the effect on sensitivity of the detection of the bio-diagnostic platform.

2. Materials and Methods

2.1. Steel Substrate Preparation

Stainless steel (316L) was used. The different steps of the process chain were optimized using steel discs (30 mm diameter, 3 mm thickness). The surfaces of the discs were mirror-polished before surface nanostructuring using SiC grit paper and an alumina suspension. The final mold insert used for the injection molding of the bio-diagnostic platform was fabricated using conventional micromilling processes. A 500 μm wide, 150 μm high ridge, corresponding to the microchannel on the plastic part, was first micromilled. An array of microholes (diameter: 300 μm, depth: 100 μm) was then fabricated on top of the ridge. The bottom of the fabricated microholes was then polished by through-mask electrochemical micromachining of the stainless steel [25]. The mirror-polished discs (also referred to as 2D substrates in the text) and microstructured steel inserts (also referred to as 3D substrates) were passivated using a nitric acid solution (20% v/v in water, 60 °C, 30 min) and thoroughly rinsed with deionized water.

2.2. Surface Nanostructuring

The surface structuring of the steel parts was carried out by first making an etch mask using nanosphere lithography and then electrochemically etching the steel. Nanosphere lithography was carried out using conditions previously described [31]. For the targeted structure size, polystyrene particles with diameter of 1 μm and 522 nm were used (Microparticles GmbH, Berlin, Germany). The particles were used as templates for the fabrication of an etch mask suitable for electrochemical etching. A metal oxide layer with a thickness of a few nanometers was deposited on the bead template and a lift-off was carried out, resulting in a nanoporous etch mask. Steel etching was carried out using conventional electrochemical dissolution conditions [26,27].

2.3. Hot Embossing and Injection Molding

Hot embossing was used to fabricate nanostructured reference surfaces for wettability and immunoassay trials. Polycarbonate (Makrolon 2207, Bayer, Leverkusen, Germany) was used. Polycarbonate was heated to 10 °C above its glass transition temperature and an embossing pressure of 1.5 MPa was used. The sample was cooled to 20 °C below its glass transition temperature before demolding.

Injection molding was used for the replication of the bio-diagnostic platform, using the material polycarbonate Makrolon 2207 (Bayer). The polymer material was dried at 120 °C for 4 h prior the injection molding. Injection molding was performed using an Engel 50 ton injection molding machine. The nozzle temperature was set at 270 °C. In order to improve the replication quality, a rapid heating and cooling process, based on highly pressurized water, was used. Water temperatures of 60 °C (low) and 170 °C (high) were set, resulting in a mold cavity temperature of 90 °C and 150 °C, respectively. The removal of trapped gas in the mold cavity was performed with a mobile vacuum system, enabling a vacuum level down to 50 mbar in the mold. The average injection pressure was 1600 bars and the cycle time was around 50 s.

2.4. Surface Characterization

Structured steel surfaces and polycarbonate replicas were characterized by atomic force microscopy (AFM Dimension Icon, Bruker, CA, USA). Tapping mode AFM was used for the characterization of topography using aluminum-coated silicon tips (typical force constant of 5 N/m) obtained from Budget sensors (Tap150Al-G, Sofia, Bulgaria). The root mean square (RMS) roughness, feature diameter/height/density and image surface area difference were all measured using the built-in functions of the NanoScope software. The image surface area difference corresponds to the difference between the image's three-dimensional surface area and its projected two-dimensional surface area. This was used to quantify the increase in specific surface area due to the presence of surface features. Scanning electron microscopy (XL-30 ESEM-FEG, Philips, The Netherlands) was used to characterize the structures produced on the 3D steel inserts and the polycarbonate replicas.

2.5. Wettability Measurements

The wettability changes of the surfaces were characterized by measuring the contact angle of water sessile droplets deposited on the sample. Advancing and receding water contact angles were determined using a Drop Shape Analysis System DSA10 provided by Krüss (Hamburg, Germany). Standard deviations were calculated using three measurements and the error bars shown on the graphs correspond to the 95% confidence intervals.

2.6. Immunoassay

The model immunoassay was carried out by first spotting mouse immunoglobulin (IgG, Jackson ImmunoResearch Europe, Newmarket, UK) using a Nano-PlotterTM (GeSIM, Großerkmannsdorf, Germany). The surface was blocked using bovine serum albumin (BSA) to prevent non-specific adsorption (Jackson ImmunoResearch Europe). The excess BSA was then removed by rinsing with phosphate-buffered saline (PBS) with a surfactant (Tween20) and then PBS only. PBS and PBS-Tween were supplied by Sigma Aldrich (Buchs, Switzerland). The complementary antibody (αIgG, Jackson ImmunoResearch Europe) conjugated to a fluorescent marker (Cy5) was then added. After a last rinsing step (PBS/Tween 20 and PBS) to remove any αIgG excess, a confocal microscope (TCS-SP5, Leica, Heerbrug, Switzerland) was used to image the detection spots. The amount of fluorescence was characterized by measuring the camera-gain necessary to barely reach saturation.

3. Results and Discussion

3.1. Surface Structuring of Flat Stainless Steel (2D Substrate)

The first part of this study focused on the fabrication of sub-micro-/nanostructures on flat, stainless steel substrates.

As shown in Figure 1, the proposed process flow consists of four main steps: first, the deposition of micro-/nanoparticles on the surface of the steel; second, the deposition of a thin (a few nanometers thick) metal oxide layer; third, a lift-off step to remove the particles and produce a porous metal oxide etch mask. Finally, the structures are transferred into the substrate by etching the stainless steel by means of an electrochemical dissolution process. The lateral dimensions are controlled via the particle deposition step (particle diameter, particle density). The vertical dimensions are given by the etching step (depth, profile).

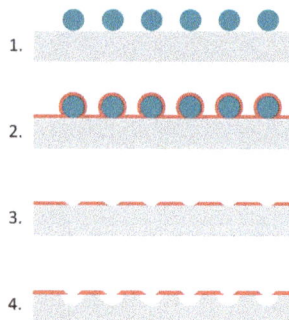

Figure 1. Schematic showing the process flow used for steel nanostructuring.

In Figure 2, photographs and SEM images of flat stainless steel (316L) surfaces coated with particles are presented. The presence of the particles leads to a color change effect due to light scattering. These color changes depend on the size of the particles and the viewing angle and are well described [31]. For 500 nm particles, a monolayer was obtained with a particle density of 1.49×10^8 part·cm^{-2} and a fill factor of 33%. For 1 μm particles, a density of 2×10^7 part·cm^{-2} was obtained with a fill factor of 17%. The particle monolayer was then used for the fabrication of an etch mask before the electrochemical dissolution of the steel. The pattern was transferred into a thin hard mask prior to the electrochemical dissolution. Figure 3 presents the result obtained using 500 nm beads as templates.

The samples were then electrochemically etched. In Figure 4, a photograph of the steel surface after etching and the corresponding AFM image of the topography are presented. As for the particle-coated surfaces, the transfer of the structures into the steel gave a color change effect on the surface. AFM confirmed that hemispherical structures were obtained. For an etch mask made using a template of 500 nm beads, the average diameter of the final structures after etching was 460 nm with an average depth of 260 nm. One of the main advantage of electrochemical dissolution is the surface quality obtained in the etched areas, which leads to an RMS roughness on the order of a few nanometers [26]. A range of structures has been fabricated on these flat reference steel substrates using different particles as templates and different etching conditions. Hemispherical structures with diameters of 460 nm and 950 nm have been produced with aspect ratios of up to 1:2.

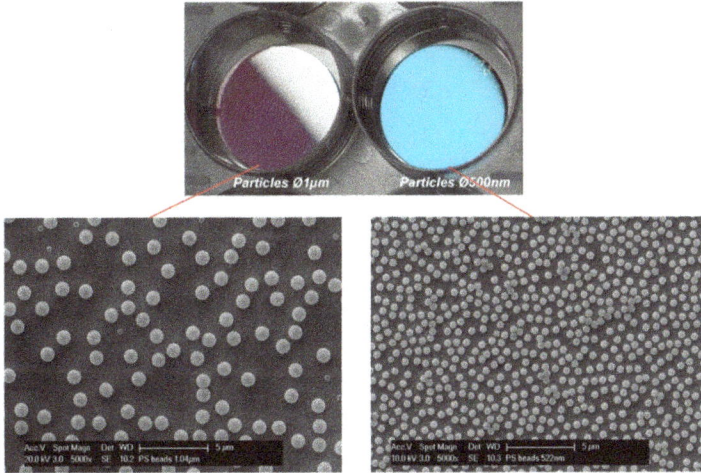

Figure 2. (**Top**) photograph of flat stainless steel inserts coated with 1 μm and 500 nm particles. Bottom (**left**) SEM images of the 1 μm particle monolayer. Bottom (**right**) SEM images of the 500 nm particle monolayer.

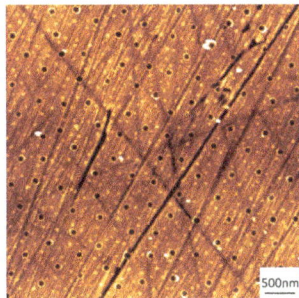

Figure 3. Atomic force microscopy (AFM) image of the stainless steel surface after the deposition of the etch mask.

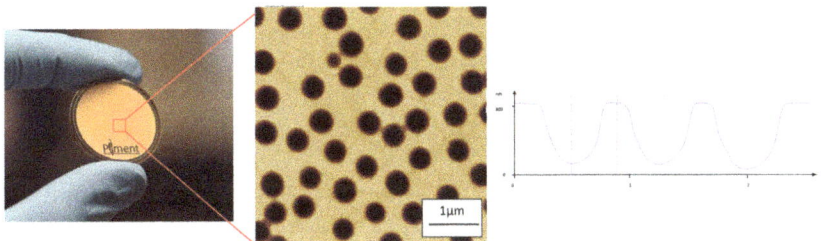

Figure 4. Photograph and AFM images of a flat stainless steel inserts after electrochemical etching.

3.2. Surface Nanostructuring of the Microstructured Insert (3D Substrate)

The second part of this study focused on the fabrication of sub-micro-/nanostructures on top of the detection spots of an injection-molded bio-diagnostic platform. To this end, the microstructured steel insert used for injection molding had to be processed. In Figure 5, a photograph of the stainless steel insert fabricated by micromilling is presented. The insert had four mounting-holes for its integration in the mold, holes for the ejector and a micro-ridge corresponding to the microchannel of the bio-diagnostic platform. The array of microholes is clearly visible on top of the ridge. The bottoms of the microholes were electropolished after micromilling to reduce their surface roughness.

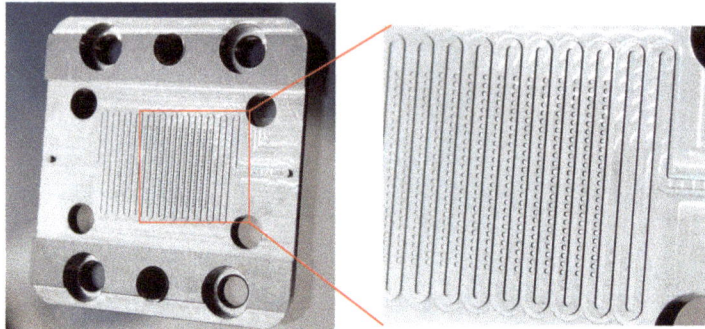

Figure 5. Photograph of the stainless steel insert of the bio-diagnostic platform made by micromilling.

The same process as was used for planar substrates was applied to the insert for the deposition of particles (500 nm diameter). As shown in Figure 6, a homogeneous deposition is obtained on the insert. Particles are observed at the bottom of the holes and on top of the ridge. The background roughness does not influence the particle deposition process and particles can also be observed on the side walls of the microholes (data not shown). The particle density at the bottom was 1.43×10^8 part·cm^{-2} with a fill factor of 25%. This is comparable to the results obtained on a flat surface.

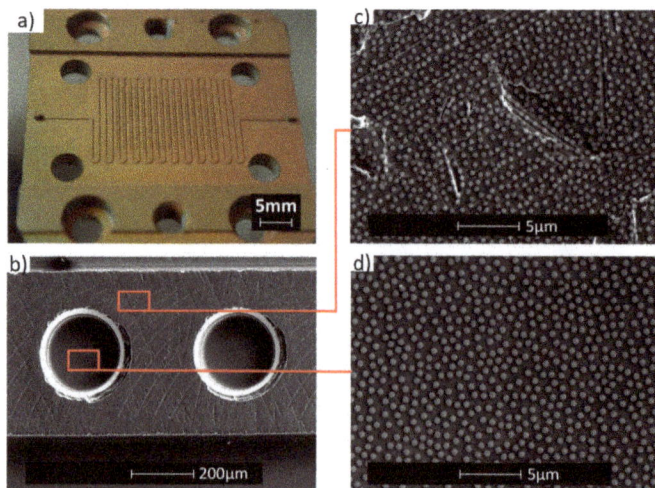

Figure 6. Photograph (**a**) and SEM images (**b–d**) of the stainless steel insert of the bio-diagnostic platform coated with particles.

After the fabrication of the etch mask and electrochemical etching, sub-micro-/nanoholes were observed at the bottom of the microholes (Figure 7). SEM was used to qualitatively examine the structures created at the bottom of the holes. AFM could not be used to make the surface characterization within the microholes. A combination of the high aspect ratio of the holes and the geometry of the AFM cantilever did not allow engagement of the tip at the bottom of the microhole.

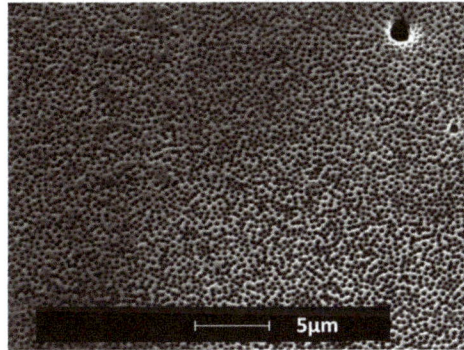

Figure 7. SEM image of the bottom of a microhole of the insert after electrochemical etching.

3.3. Hot Embossing and Injection Molding of Plastic Parts

The structured 2D insert and the bio-diagnostic platform insert were used as molds for replication into a single thermoplast (polycarbonate Makrolon 2207). Hot embossing was used for the nanostructured 2D substrate and injection molding was used for the bio-diagnostic platform. Figure 8 presents the AFM characterization of a hot embossed replica. As expected, sub-micro hemispherical bumps were obtained, the height of which corresponded to the depth of the holes fabricated in the insert. However, the AFM sections show that, while the smallest structures were easily demolded, the largest structures showed side wall defects due to issues with the demolding.

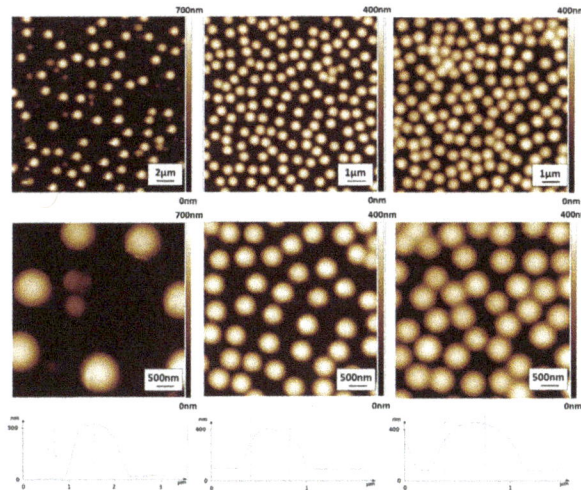

Figure 8. AFM images of the replica hot embossed using three different structured inserts. (**left**): structure 1; (**middle**): structure 2; (**right**): structure 3.

In the case of the bio-diagnostic platform, nanostructures were injection-molded. Figure 9 presents an SEM image of the structures obtained on top of the microdetection spot. The nanostructures were successfully replicated in the polycarbonate.

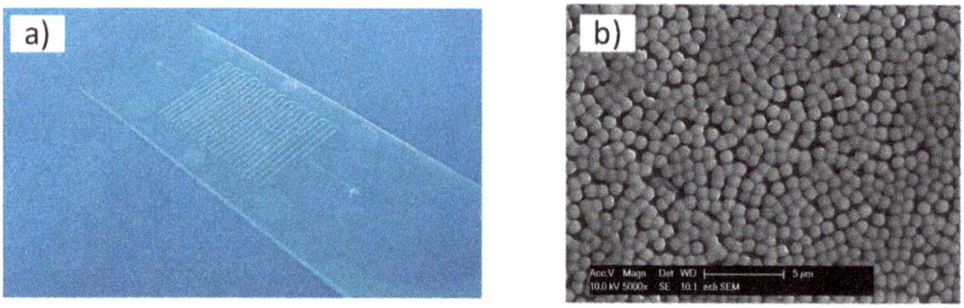

Figure 9. (**a**) Photograph of the injection-molded bio-diagnostic platform. (**b**) SEM image of the nanostructures on the spot of the injection-molded bio-diagnostic platform.

However, due to characterization limitations, a quantitative assessment of the replication is impossible. The nanostructured surface to be characterized (the top of the microdetection spots) on the polycarbonate replica could not be reached by the AFM tip due to its location within the microchannel of the device.

3.4. Characterization

Two types of characterization were carried out. First, the wettability of the samples produced by hot embossing was characterized by measuring water contact angles. Second, a complete immunoassay was performed on all samples to investigate the effect of surface structuring on the sensitivity of the bio-diagnostic platform.

3.4.1. Surface Characterization

Table 1 presents the results of the AFM characterization. The RMS roughness, surface area difference, feature diameter, feature height and feature density have been measured for all three structures from the AFM images. The surface area difference of structure 1 is only 14.7%, whereas it is above 30% for structures 2 and 3. This can be explained by the low feature density of structure 1, which is an order of magnitude lower than that of structures 2 and 3. When we compared the three structures in terms of feature density, we found that structure 1 has the lowest feature density, followed by structures 2 and 3, which have similar values. Concerning the diameter and height of the feature, structure 1 has the largest, followed by structure 3 and then structure 2. These data show that we have three different cases; a low feature density with large structures (structure 1), a high feature density with small structures (structure 2) and finally a high feature density with large structures (structure 3).

Table 1. Root mean square (RMS) roughness, surface area difference, average feature diameter and height of the three structures.

ID	RMS Roughness (nm)	Surface Area Difference (%)	Feature Density (part/cm^2)	Average Feature Diameter (nm)	Average Feature Height (nm)
Structure 1	143	14.7	1.5e7	1249	516
Structure 2	107	33.3	1.37e8	753	287
Structure 3	121	37.3	1.34e8	914	360

3.4.2. Wettability

Polycarbonate samples with three different structures were tested. A flat polycarbonate surface was also used as a reference. Wettability is characterized by measuring advancing and receding contact angles of water used as probe-liquid. The results are presented in Figure 10.

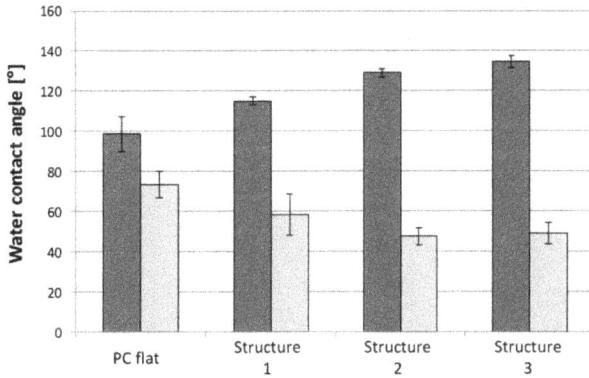

Figure 10. Water contact angles: advancing (dark grey) and receding (light grey) measured on polycarbonate with four different types of structures.

For the flat reference, an advancing contact angle of almost 100° was observed with a wetting hysteresis of 30°. The surface micro-/nanostructures lead to not only an increase in the advancing contact angle, but also a significant increase in the contact angle hysteresis for all structures tested. For structure 1, an advancing contact angle of 115° and a hysteresis of 57° were measured. For structures 2 and 3, an advancing contact angle of 135° and a hysteresis of 85° is obtained. This increase in hysteresis suggests that the water drops are in the Wenzel mode (complete wetting of the structure), which corresponds to an increase in the adhesion of the drop on the surface [5]. One objective of the surface structuring is to control the wetting of the solution during the spotting step of the immunoassay. To characterize this, a solution of fluorescently labelled proteins was inkjet printed onto flat and structured polycarbonate surfaces. No significant difference was observed between the flat control and the structured samples.

3.4.3. Immunoassay

The effect of surface micro-/nanostructuring on the sensitivity of a standard immunoassay was investigated. The protocol was applied to either the micro-/nanostructured hot-embossed surfaces or to the detection spots of the injection-molded bio-diagnostic platforms. Fluorescence microscopy was used to characterize the homogeneity of the spots. The sensitivity of the immunoassay was characterized by measuring the gain necessary to reach saturation on the camera. The results are presented in Figure 11. The quantification of the fluorescence signal revealed that the presence of structure 3 caused an increase in the sensitivity of the immunoassay. Compared to the flat reference, the gain necessary to reach saturation was lowered by 30%. The hypothesis initially proposed to explain this effect is that it is due to the increase in the specific surface area resulting from the surface structures. However, the AFM characterization of the structured surfaces revealed 15%, 33% and 37% increases in specific surface area for structures 1, 2 and 3, respectively. Although they have similar specific surface areas, structure 2 and structure 3 did not lead to a similar increase in fluorescence intensity. A second hypothesis is that variation in surface roughness has an effect on the adsorption and conformation of proteins on the surfaces. As shown by Scoppeliti et al., surface roughness can significantly affect the adsorption of proteins during immunoassays and can lead to increased protein density on the surface, which may increase the sensitivity of the immunoassay [32]. A third hypothesis is that the interaction of light with the surface structures may influence the fluorescence emission level. Highly sensitive sensors have been produced using optical interference to improve fluorescence signal [33]. In a previous publication, we showed that particle monolayers deposited on a surface lead to significant optical effects due to the interference between the incident light and the light scattered by the particles [31]. This size-dependent effect is also expected to have occurred with the hemispherical structures of this study. One other hypothesis, therefore, is that the signal enhancement is due to an optical interference effect caused by the presence of the structures. A more detailed analysis of the optical properties of the structured surfaces in the immunoassay medium is needed to confirm this hypothesis and the origin of this enhancement.

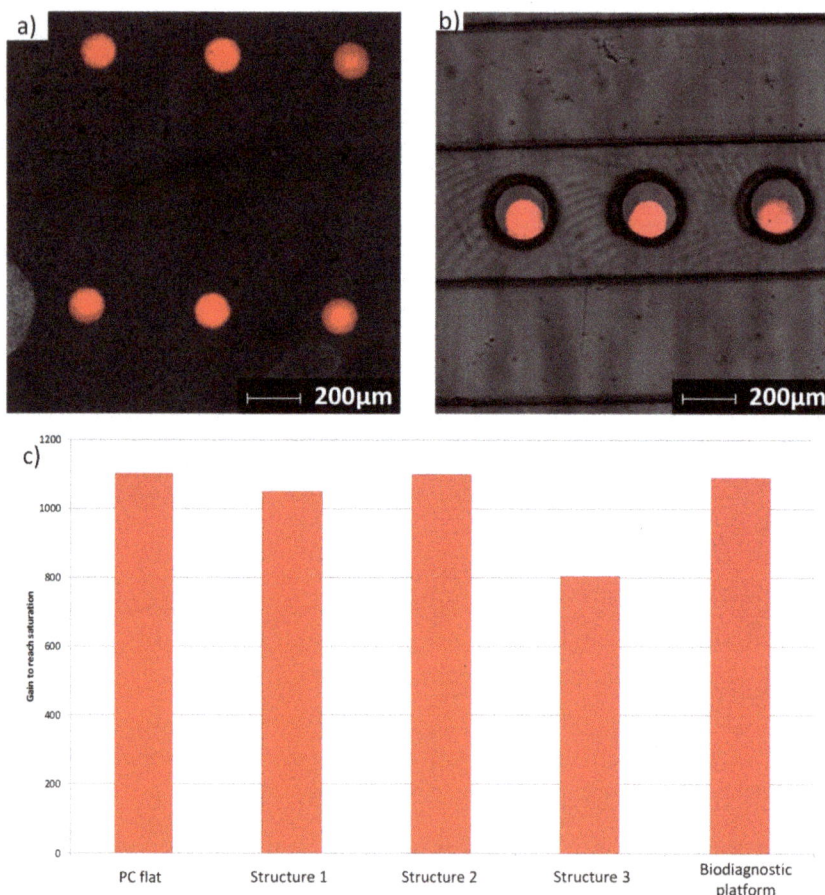

Figure 11. Confocal images of the fluorescent spots printed on a hot embossed sample (**a**) and injection-molded samples (**b**). (**c**) Graph presenting the gains necessary to reach saturation on the fluorescent camera.

4. Conclusions

The fabrication of micro-/nanostructured steel surfaces has been achieved by combining nanosphere lithography and electrochemical etching. Structures with lateral sizes of 400 nm to 1 μm with an aspect ratio of 1:2 were produced. The process was applied to planar substrates as well as micromilled inserts presenting micro-ridges and microholes. The material used was stainless steel and it is planned to extend this approach to the structuring of tool-steel used for molds. Polycarbonate replicas were produced by hot embossing or injection molding. The wettability of the surfaces was influenced by the surface structures and an increase in the adhesion of water drops was observed (drops adopted the Wenzel wetting state). One of the structures was also found to significantly increase the sensitivity of an immunoassay, with a 30% increase in fluorescence signal.

Acknowledgments: The Interreg IVa project PIMENT is funded by the European Regional Development Fund, the Federal Office for Education and Science (OFES) and on the Swiss side by the cantons of Neuchâtel and Vaud. We thank them for their support.

Author Contributions: P.V. conceived, designed and manufactured the microstructured mold-inserts. N.B. and R.P. conceived and designed the nanostructuring process flow; N.B. performed the surface nanostructuring process. P.-F.C. and M.D. conceived, designed and performed the electrochemical polishing and etching of the mold inserts. L.T. and S.D. conceived, designed and performed the injection molding of plastic parts. N.B. conceived, designed and performed the hot-embossing tests and the wettability characterization. G.A. conceived, designed and performed the immunoassay experiments. N.B., L.T. and G.A. wrote the paper. N.B. revised the paper.

Conflicts of Interest: The authors declare no conflict of interest.

References

1.   Kim, S.; Jung, U.T.; Kim, S.K.; Lee, J.H.; Choi, H.S.; Kim, C.S.; Jeong, M.Y. Nanostructured Multifunctional Surface with Antireflective and Antimicrobial Characteristics. *Appl. Mater. Interfaces* **2015**, *7*, 326–331. [CrossRef] [PubMed]

2.   Weder, G.; Blondiaux, N.; Giazzon, M.; Matthey, N.; Klein, M.; Pugin, P.; Heinzelmann, H.; Liley, M. Use of Force Spectroscopy to Investigate the Adhesion of Living Adherent Cells. *Langmuir* **2010**, *26*, 8180–8186. [CrossRef] [PubMed]

3.   Roch, T.; Weihnacht, V.; Scheibe, H.J.; Roch, A.; Lasagni, A.F. Direct Laser Interference Patterning of tetrahedral amorphous carbon films for tribological applications. *Diam. Relat. Mater.* **2013**, *33*, 20–26. [CrossRef]

4.   Quéré, D. Wetting and roughness. *Annu. Rev. Mater. Res.* **2008**, *38*, 71–99. [CrossRef]

5.   Blondiaux, N.; Scolan, E.; Popa, A.M.; Gavillet, J.; Pugin, R. Fabrication of superhydrophobic surfaces with controlled topography and chemistry. *Appl. Surf. Sci.* **2009**, *256*, S46–S53. [CrossRef]

6.   Spori, D.; Drobek, T.; Zürcher, S.; Spencer, N.D. Cassie-State wetting investigated by means of holes to pillar density gradient. *Langmuir* **2010**, *26*, 9465–9473. [CrossRef] [PubMed]

7.   Martines, E.; Seunarine, K.; Morgan, H.; Gadegaard, N.; Wilkinson, C.D.W.; Riehle, M.O. Superhydrophobicity and superhydrophilicity of regular nanopatterns. *Nano Lett.* **2005**, *5*, 2097–2103. [CrossRef] [PubMed]

8.   Blondiaux, N.; Scolan, E.; Franc, G.; Pugin, R. Manufacturing of super-hydrophobic surfaces combining nanosphere lithography with replication techniques. *Nanotechnology* **2012**, 1–6. [CrossRef]

9.   Krishnamoorthy, S. Nanostructured sensors for biomedical applications—A current perspective. *Curr. Opin. Biotechnol.* **2015**, *34*, 118–124. [CrossRef] [PubMed]

10. Stewart, M.E.; Anderton, C.R.; Thompson, L.B.; Maria, J.; Gray, S.K.; Rogers, J.A.; Nuzzo, R.G. Nanostructured plasmonic sensors. *Chem. Rev.* **2008**, *108*, 494–521. [CrossRef] [PubMed]

11. Ko, H.; Singamaneni, S.; Tsukruk, V.V. Nanostructured surfaces and assemblies as SERS media. *Small* **2008**, *4*, 1576–1599. [CrossRef] [PubMed]

12. Kanipe, K.N.; Chidester, P.P.F.; Stucky, G.D.; Moskovits, M. Large format surface-enhanced Raman spectroscopy substrate optimized for enhancement and uniformity. *ACS Nano* **2016**, *10*, 7566–7571. [CrossRef] [PubMed]

13. Ingham, C.J.; Maat, J.; De Vos, W.M. Where bio meets nano: The many uses for nanoporous aluminum oxide in biotechnology. *Biotechnol. Adv.* **2012**, *30*, 1089–1099. [CrossRef] [PubMed]

14. Kim, J.S.; Chob, J.B.; Park, B.G.; Leed, W.; Lee, K.B.; Oh, M.K. Size-controllable quartz nanostructure for signal enhancement of DNA chip. *Biosens. Bioelectron.* **2011**, *26*, 2085–2089. [CrossRef] [PubMed]

15. Kuwabara, K.; Ogino, M.; Ando, T.; Miyauchi, A. Enhancement of fluorescence intensity from an immunoassay chip using high-aspect-ratio nanopillars fabricated by nanoimprinting. *Appl. Phys. Lett.* **2008**, *93*, 3. [CrossRef]
16. Masuzawa, T. State of the Art of Micromachining. *Ann. CIRP* **2000**, *49*, 473–488. [CrossRef]
17. Bieda, M.; Schmädicke, C.; Roch, T.; Lasagni, A. Ultra-low friction on 100Cr6-steel surfaces after direct laser interference patterning. *Adv. Eng. Mater.* **2015**, *17*, 102–108. [CrossRef]
18. Madou, M. *Fundamentals of Microfabrication: The Science of Miniaturization*; CRC Press: Boca Raton, FL, USA, 2002.
19. Pease, R.F.; Chou, S.Y. Lithography and other patterning techniques for future electronics. *Proc. IEEE* **2008**, *96*, 248–270. [CrossRef]
20. Ruiz, R.; Kang, H.; Detcheverry, F.A.; Dobisz, E.; Kercher, D.S.; Albrecht, T.R.; De Pablo, J.J.; Nealey, P.F. Density multiplication and improved lithography by directed block copolymer assembly. *Science* **2008**, *321*, 936–939. [CrossRef] [PubMed]
21. Klein, M.J.K.; Montagne, F.; Blondiaux, N.; Vazquez-Mena, O.; Heinzelmann, H.; Pugin, R.; Brugger, J.; Savu, V. SiN membranes with submicrometer hole arrays patterned by wafer-scale nanosphere lithography. *J. Vac. Sci. Technol. B* **2011**, *2*, 021012. [CrossRef]
22. Hao, L.; Meng, Y.; Chen, C. Experimental investigation on effects of surface texturing on lubrication of initial line contacts. *Lubr. Sci.* **2014**, *26*, 363–373. [CrossRef]
23. Mason, J.A.; Adams, D.C.; Johnson, Z.; Smith, S.; Davis, A.W.; Wasserman, D. Selective thermal emission from patterned steel. *Opt. Express* **2010**, *18*, 25192. [CrossRef] [PubMed]
24. Rao, P.N.; Kunzru, D. Fabrication of microchannels on stainless steel by wet chemical etching. *J. Micromech. Microeng.* **2007**, *17*, 99–106.
25. Landolt, D.; Chauvy, P.F.; Zinger, O. Electrochemical micromachining, polishing and surface structuring of metals: Fundamental aspects and new developments. *Electrochim. Acta* **2003**, *48*, 3185–3201. [CrossRef]
26. Zinger, O.; Chauvy, P.F.; Landolt, D. Development of titatnium electrochemical microstructuring towards implant applications. *Eur. Cells Mater.* **2001**, *1*, 24–25.
27. Zinger, O.; Chauvy, P.F.; Landolt, D. Scale-resolved electrochemical surface structuring of titanium for biological applications. *J. Electrochem. Soc.* **2003**, *150*, B495–B503. [CrossRef]
28. Shimizu, M.; Yamada, T.; Sasaki, K.; Takada, A.; Nomura, H.; Iguchi, F.; Yugami, H. Anisotropic multi-step etching for large-area fabrication of surface microstructures on stainless steel to control thermal radiation. *Sci. Technol. Adv. Mater.* **2015**, *16*, 025001–025006. [CrossRef] [PubMed]
29. Al-Azawi1, A.; Smistrup, K.; Kristensen, A. Nanostructuring steel for injection molding tools. *J. Micromech. Microeng.* **2014**, *24*, 055023–055029. [CrossRef]
30. Kurihara, K.; Saitou, Y.; Souma, N.; Makihara, S.; Kato, H.; Nakano, T. Fabrication of nano-structure anti-reflective lens using platinum nanoparticles in injection moulding. *Mater. Res. Express* **2015**, *2*, 015008. [CrossRef]
31. Guillaumée, M.; Liley, M.; Pugin, R.; Stanley, R. Scattering of light by a single layer of randomly packed dielectric microspheres giving color effects in transmission. *Opt. Express* **2008**, *16*, 1440–1447. [CrossRef] [PubMed]
32. Scopelliti, P.E.; Borgonovo, A.; Indrieri, M.; Indrieri, M.; Giorgetti, L.; Bongiorno, G.; Carbone, R.; Podesta, A.; Milani, P. The Effect of Surface Nanometre-Scale Morphology on Protein Adsorption. *PLoS ONE* **2010**, *5*, 11862. [CrossRef] [PubMed]
33. Yasuda, M.; Akimoto, T. Highly Sensitive Fluorescence Detection of Avidin/Streptavidin with an Optical Interference Mirror Slide. *Anal. Sci.* **2012**, *28*, 947–952. [CrossRef] [PubMed]

micromachines

MDPI

Article

Fabrication of Mesoscale Channel by Scanning Micro Electrochemical Flow Cell (SMEFC)

Cheng Guo, Jun Qian * and Dominiek Reynaerts

Department of Mechanical Engineering, KU Leuven & Member Flanders Make, Leuven 3001, Belgium;
cheng.guo@kuleuven.be (C.G.); dominiek.reynaerts@kuleuven.be (D.R.)
* Correspondence: jun.qian@kuleuven.be; Tel.: +32-16-322-524

Academic Editors: Guido Tosello and Nam-Trung Nguyen
Received: 6 March 2017; Accepted: 27 April 2017; Published: 4 May 2017

Abstract: A unique micro electrochemical machining (ECM) method based on a scanning micro electrochemical flow cell (SMEFC), in which the electrolyte is confined beneath the tool electrode instead of spreading on the workpiece surface, has been developed and its feasibility for fabricating mesoscale channels has been investigated. The effects of the surface conditions, the applied current, the feed rate, the concentration of the electrolyte and several geometrical parameters on the machining performance have been investigated through a series of experiments. The cross-sectional profile of the channels, the roughness of the channel bottom, the width and depth of the channel, the microstructures on the machined surface and the morphologies of the moving droplet have been analyzed and compared under different machining conditions. Furthermore, experiments with different overlaps of the electrolyte droplet traces have also been conducted, in which the SMEFC acts as a "milling tool". The influences of the electrode offset distance (EOD), the current and the feed rate on the machining performance have also been examined through the comparison of the corresponding cross-sectional profiles and microstructures. The results indicate that, in addition to machining individual channels, the SMEFC system is also capable of generating shallow cavities with a suitable superimposed motion of the tool electrode.

Keywords: electrochemical machining (ECM); scanning micro electrochemical flow cell (SMEFC); micro-ECM; channel machining

1. Introduction

Nowadays, there is an increasing demand of mesoscale channel structures in many industrial domains such as fuel cells [1], hydrodynamics bearing [2], and sealing ring channel [3]. Electrochemical machining (ECM) has proven to be a unique method for fabricating channels, ranging from the micro scale to the macro scale, with excellent surface integrities on metallic materials. Ryu [4] utilized a micro foil electrode instead of a micro-shaft electrode to achieve micro grooving in the environment of the citric acid electrolyte and a 42 μm wide and 18 μm deep single channel with 11 nm Ra surface roughness was obtained. Liu et al. [5] has utilized a kind of ECM method with low-frequency vibrations to fabricate multiple slots for the application of fuel cells. The same features were also successfully fabricated by other researchers [6], in which they developed an innovative multifunctional cathode, combining the tool electrode, the sealing device and the spacers between every channels. Electrolyte jet machining (jet-ECM) is also a promising technique for channel machining. Natsu et al. [7] machined complex channels on a thrust hydrodynamic bearing surface with jet-ECM in an efficient way. Compressed air assisted jet-ECM was also attempted to fabricate channels on the workpiece made of Nimonic Alloy 80a [8]. Instead of using conventional jet-ECM with a round nozzle scanning on the workpiece to fabricate channels [9], Kunieda et al. [10] utilized a flat nozzle to shorten the channel machining time. With the same method, Hackert-Oschätzchen et al. [11] fabricated complex channels with a width below 200 μm for the application of microfluidics and micro reactors. Apart from these methods, electrochemical milling has also been verified as a feasible method for channel machining. As an example, Ghoshal and Bhattacharyya [12] used micro tools with different front end shape for microchannel machining with the scanning machining layer and layer method. Furthermore, they investigated the optimum scan feed rate in electrochemical micromachining of micro channels and found that the introduction of the optimum scan feed rate can reduce overcut and avoid the breakage of micro tools during the grooving process [13]. Besides, Kim et al. [14] utilized the electrode with a diameter of 38 μm to fabricate a 50 μm wide micro channel thanks to ultrashort pulses. Zhang et al. [15] also used an electrode with the diameter of 10 μm to machine micro channels with the width of 20 μm. For machining micro through-grooves, the micro wire electrode electrochemical cutting method is also a promising way. Shin et al. [16] used a 10 μm diameter tungsten wire as the tool electrode and obtained micro channel of around 20 μm in width on stainless steel. Wang et al. [17] obtained multi-microchannels by applying low frequency and small amplitude vibration on the wire electrode based on the conventional setup. Recently, hybrid processes such as ECM with slurry jet [18], laser-assisted jet-ECM [19] and laser-assisted ECM milling [20] have brought in new possibilities for the efficient fabrication of channels on hard metals with enhanced machining localization.

Currently, the growing environment awareness in the research community has brought forth a trend for clean manufacturing in the domain of electro-physical and chemical machining. Ryu [4] has proposed a micro electrochemical reverse drilling method, in which the electrochemical reactions are confined in a droplet formed at the bottom edge of the workpiece. The droplet can be stabilized by the surface tension and the gravity, keeping other regions untouched. Sakairi et al. have developed a co-axial dual capillary solution flow type droplet cell to accomplish Ni deposition [21]. Unlike in the traditional case, the deposition happens in a container filled with electrolyte. Drensler et al. [22] proposed a method to use scanning droplet cell to selectively dissolve NiAl matrix and release the embedded W nanowires, which is intended for the application of self-assembly. Hu et al. [23] also proposed a gushing and sucking method with a coaxial tube to achieve the groove width of 103 μm and the surface roughness of 0.012–0.025 μm. Kuo et al. [24] tried to process quartz glass by electrochemical discharge machining (ECDM) with titrated flow of electrolyte, leading to less cost and pollution because of the electrolyte supplied in droplets. These methods have some special advantages compared with the conventional ones, such as keeping non-processing region untouched, better safety for operators and more feasibility to be integrated into other process chains.

In this research work, a scanning micro electrochemical flow cell (SMEFC) has been proposed to generate channels on metallic workpieces. In the SMEFC, the electrolyte is confined in a small droplet and its refreshing is simultaneously maintained. In this way, electrolyte splashing does not exist, so there is no need of an electrolyte tank for the machining region. As a result, this technique can be conveniently integrated into other manufacturing process because of its unique control of the electrolyte. The influence of surface condition on the machining performance was investigated firstly. Then, several process parameters (e.g., the current, the feed rate and the concentration of the electrolyte) and geometrical parameters have also been varied to investigate the effects on the channel formation process, in terms of dimensional parameters and surface microstructures. After analyzing the machining of single channels by SMEFC, superimposed process of SMEFC with different lateral overlapping has also been examined to study the feasibility of SMEFC for electrochemical milling.

2. SMEFC Experimental System

Figure 1 shows the schematics of the SMEFC system. The principle of the electrolyte circulation is that the electrolyte is pumped through a hollow electrode and then it arises along the electrode outer wall by the surrounding flowing air induced by the Venturi effect, resulting in a droplet between the electrode and the workpiece. This method maintains the electrolyte of the droplet fresh and confines the electrolyte in the region of interest. The used electrolyte with the reaction products flows eventually into the waste electrolyte tank through a channel in the suction head. The vacuum gap (VG) and the inter-electrode gap (IEG) can be adjusted.

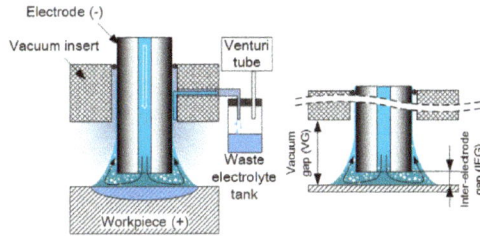

Figure 1. Schematics of the scanning micro electrochemical flow cell (SMEFC).

The solid model of the experimental setup in Figure 2 depicts the tool electrode is positioned in the collet. The suction head can move up and down through manually tuning a stage. The stage can also adjust the hole of the suction head to the center of the electrode. A flexible membrane is used to seal the suction head for the recycling of the electrolyte. The hole diameter in the suction head is 1 mm. The electrode is made of tungsten carbide, with an outer diameter of 0.5 mm and an inner diameter of 0.18 mm. The electrode is glued with a flexible tube, through which the electrolyte is pumped, as shown in Figure 2. The actual layout of the suction head, workpiece and the electrode wrapped with electrolyte are shown in Figure 2.

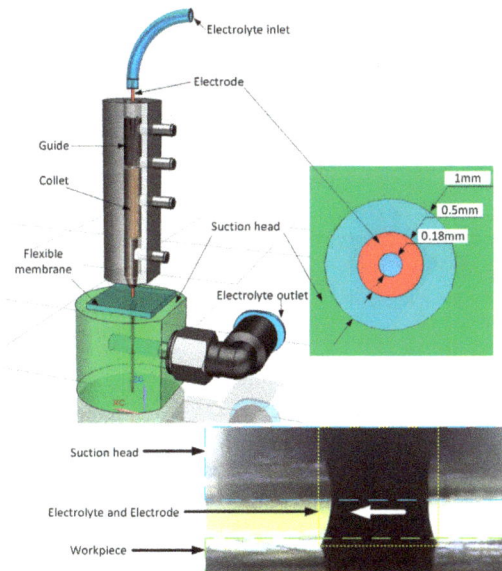

Figure 2. Layout of SMEFC setup.

The ECM power supply is a homemade switching power generator, working at 150 kHz. This generator can be set to either constant-current or constant-voltage working mode. The output ripple is less than 20 mV and the response time is less than 50 ms. The motorized stage is MTS25-Z8 of THORLABS with the travel range of 25 mm and the maximum velocity of 2.4 mm/s. The minimum achievable incremental movement is 0.05 μm and the bidirectional repeatability is 1.6 μm.

The cross-sectional profiles and the roughness of the channels machined by SMEFC have been measured by a Mitutoyo CS3200 profiler (Mitutoyo, Kawasaki, Japan). When measuring the roughness, the workpiece is cleaned firstly in a mixture of ethanol and acetone in an ultrasonic tank. The stylus contacts the bottom surface of the channel and moves along the channel. The nominal radius of the stylus is 2.0 μm. A Zeiss optical microscope (Carl Zeiss AG, Oberkochen, Germany), SteREO Discovery V20, was used to evaluate the width of the channels cavities, and Dino-Lite digital microscope AM4115ZT (AnMo Electronics Corporation, Hsinchu, Taiwan) was installed to help monitor and observe the moving electrolyte droplet above the workpiece top surface. The maximum lateral resolution of the digital microscope is around 1.4 μm/pixel. The surface microstructures of the channel bottom surface were examined with a Phenom desktop SEM (Phenom-World B.V., Eindhoven, The Netherlands). The current signals during the electrochemical dissolution were recorded by a data acquisition unit embedded in the ECM power supply.

3. SMEFC Machining Experiments and Discussion

A series of experiments have been carried out under constant-current mode. The electrolyte is a sodium nitrate ($NaNO_3$) solution with the concentration of 120 g/L and 250 g/L. The workpiece material is a kind of stainless steel from UDDEHOLM STAVAX® ESR (Hagfors, Sweden). The electrolyte flow rate is 0.06 mL/s pumped by a metering pump of ProMinent® (ProMinent GmbH, Heidelberg, Germany). The vacuum condition is achieved by a Venturi tube, with an inlet pressure of 4 bar. The corresponding vacuum pressure is 45 kPa.

3.1. Experimental Verification

When the droplet moves on the workpiece, the droplet shape will be regulated by its gravity, surface tension and adhesion simultaneously. Therefore, surface conditions or the roughness of the workpiece may affect the behavior of the moving droplet. To investigate this conjecture, samples with two different surface treatments were used. One sample was ground by a surface grinder (Ra = 0.4 μm) and the other one was pre-machined by micro-EDM milling (Ra = 1.2 μm). The surface tension of the electrolyte is around 75.5 mN/m [25].

Figure 3 shows the profile of the electrolyte droplet moving at different speed on a micro-EDMed surface. The moving direction of the electrolyte droplet relative to the workpiece has been indicated by the arrows. The applied current is 230 mA and the feed rates are 0.1 mm/s, 0.2 mm/s, 0.4 mm/s and 0.8 mm/s, respectively. It can be noticed that when the scanning speed is 0.1 mm/s, the droplet seems to be not stable. The electrolyte left on the workpiece can be obviously observed and it crystallizes very quickly probably due to its exposure to surrounding flowing air. With a slower feed rate, the channel becomes deeper and the corresponding droplet volume in the channel becomes bigger in the same time. When the volume reaches a certain level, the surface tension (cohesion) cannot support the volume in the form of a single droplet. As a result, some of the electrolyte is left on the machined surface, as illustrated in Figure 4. Consequently, the electrolyte cannot be fully recovered by the Venturi effect. The electrolyte droplets above the workpiece top surface under the other three feed rates maintain a relatively stable shape during the motion. However, when the feed rate goes even higher, the cross-sectional profile perpendicular to the moving direction becomes increasingly asymmetrical.

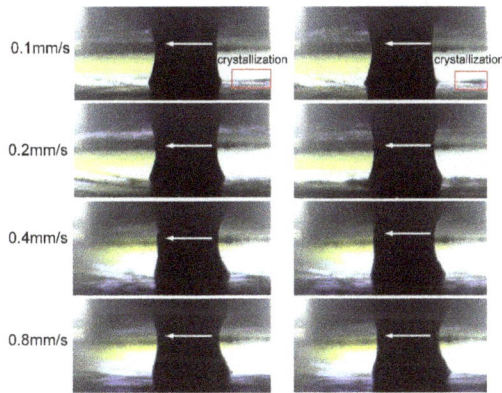

Figure 3. Electrolyte droplet moving on the surface treated by micro-electrical-discharge machining (EDM) milling.

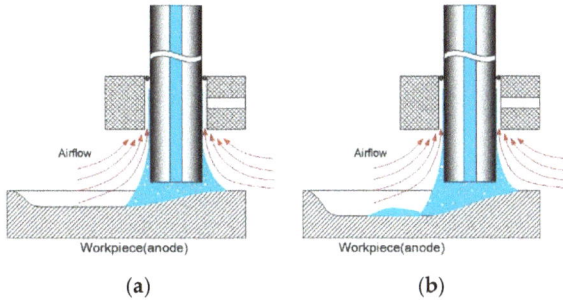

(a)   **(b)**

Figure 4. Explanation of the electrolyte leakage. (**a**) Surface tension maintains a single droplet; (**b**) electrolyte left on machined surface once the droplet volume is larger.

Figure 5 depicts the situation when the electrolyte droplet moves on the ground surface with the feed rates of 0.1 mm/s, 0.2 mm/s and 0.4 mm/s, respectively. Obvious delaying of the electrolyte droplet can be noticed. Compared with the case in Figure 3, the trail above the workpiece top surface becomes longer. Obviously, this difference in the droplet morphology is induced by the difference in the roughness of these two samples.

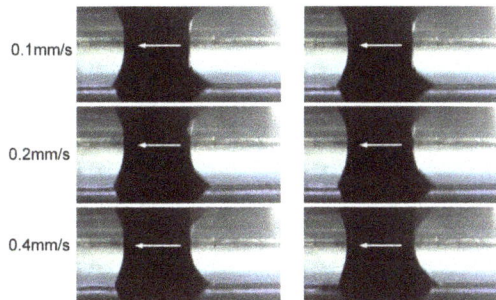

Figure 5. Electrolyte droplet moving on the ground surface.

Comparing the droplet profiles in Figures 4 and 5, it can be concluded that the surface roughness indeed influences the morphology of the moving droplet and further affects the flow field and the electric field distribution. After this initial stage of comparison, only ground samples were used in the further experiments.

3.2. Effects of Current Density and Feed Rate

Different current settings, i.e., 100 mA, 200 mA and 300 mA, have been, respectively, applied in the channel machining under a variety of feed rate of 0.1 mm/s, 0.2 mm/s, 0.3 mm/s and 0.4 mm/s. The vacuum gap (VG) and the inter-electrode gap (IEG) were set to 200 μm and 50 μm respectively. When the feed rate is set to 0.1 mm/s, relatively large fluctuations in the current signals take place around 300 mA (Figure 6), in comparison to the cases under the other three feed rates. It also means the flow field under this feed rate is unstable. Therefore, the electricity consumption per unit length (ECPL) is defined in this research and its value should be limited in order to obtain a stable flow field and to avoid low field instability.

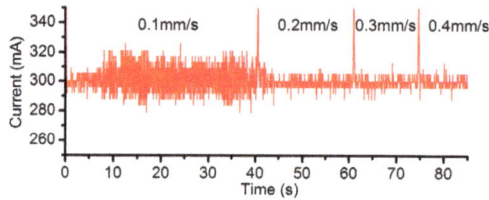

Figure 6. Current signal when setting 300 mA as the target.

In order to evaluate the effects of the feed rate and the current on the grooving performance, the ECPL was first set at a constant value in the experiments. Then, the values of current and the feed rate were varied within a range according to this rule. The first three columns of Table 1 list the five groups of parameters utilized in the experiments, with the ECPL being 1 C/mm and 0.5 C/mm, respectively.

Table 1. Experimental parameters.

Current (mA)	Feed Rate (mm/s)	ECPL (C/mm)	S_{exp} (mm^2)	S_{the} (mm^2)	η_1	η_2
100	0.1	1	0.0234	0.02327	100.6%	82.6%
200	0.2	1	0.0246	0.02327	105.7%	86.8%
300	0.3	1	0.0255	0.02327	109.6%	90.0%
100	0.2	0.5	0.0110	0.01165	94.4%	77.5%
200	0.4	0.5	0.0127	0.01165	109.0%	89.5%

The experimental results were compared with the theoretical results derived by Faraday's law. Faraday's law can be summarized as,

$$m_{the} = \frac{M \times Q}{F \times n} \tag{1}$$

where m_{the} is the mass of the substance, Q the total electric charge passed through the substance, $F = 96,485$ C/mol the Faraday constant, M the molar mass of the substance and n the valency number of the ions of the substance.

After substituting Q with ECPL, the theoretical cross-sectional removal area S_{the} by SMEFC can be written as,

$$S_{the} = \frac{M \times \text{ECPL}}{F \times n \times \rho} \tag{2}$$

where S_{the} is the theoretical cross-sectional area and ρ is the density of the substance.

The current efficiency can be described as,

$$\eta = \frac{m_{exp}}{m_{the}} = \frac{S_{exp}}{S_{the}} \qquad (3)$$

where m_{exp} and S_{exp} are the experimental removal mass and experimental cross-sectional area, respectively. The valency number n of stainless steel is controversial, because it depends on the proportion of the generated Fe^{2+} ions and Fe^{3+} ions. According to the results of [26], when $Fe^{2+}:Fe^{3+} = 1:2$, $n_1 = 3.1410$. Using this value, the current efficiency η is calculated to be always larger than 100%. Similar results are also shown in [26], and the uncertain proportion of Fe^{2+} ions and Fe^{3+} ions induce this phenomenon. Therefore, $n_2 = 2.5786$ is also used in the calculations, which is obtained under the assumption that only Fe^{2+} ions are generated. η_1 and η_2 represent the current efficiency calculated under n_1 and n_2 respectively. The actual current efficiency should be a value between these two efficiencies. S_{the} is calculated under the valency number of n_2.

Figure 7 shows the cross-sectional profiles of the channels machined with the parameters in Table 1. It can be observed that when the same ECPL, the maximum depth of these channels is almost the same in spite of some slight differences (Figure 7). However, as shown in Figure 8, the width of the channels has clear deviations. At the ECPL of 1 C/mm, the channel width processed by 0.1 mm/s is the smallest and almost the same width is produced in the cases of 0.2 mm/s and 0.3 mm/s. Similarly, at the ECPL of 0.5 C/mm, the channel width increases with the feed rate. A high concentration (250 g/L) alleviates the trend. There are two possible reasons accountable for this trend. One reason is that a high current can derive a high current efficiency compared to a low current, which can be confirmed in the current efficiency in Table 1. The other possible reason is that the contact angle of the electrolyte droplet is also influenced by the applied voltage. This implies that if a relative voltage is applied, the electrolyte droplet needs to expand to a bigger area to obtain a smaller contact angle.

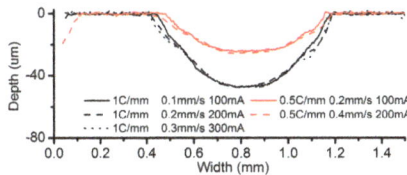

Figure 7. Cross-sectional profiles.

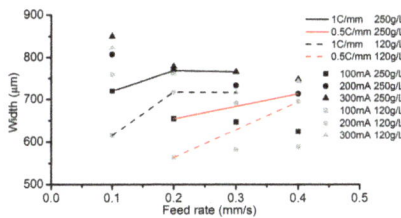

Figure 8. Channel width comparison.

Figure 9 indicates that, although the same ECPL can induce relatively uniform depth, the roughness declines along with the increase of the feed rate. Surface microstructures at the channel bottom as shown in Figure 10 also confirm this point when comparing Figure 10a,f,k. Rosenkranz et al. [27] has already pointed out, that during the ECM process, the iron surface is covered by a thin oxide layer, on top of which a polishing film of supersaturated iron nitrate forms. The thickness of film varies with the current density and the electrolyte flow rate, when other parameters do not change. With the increase of the applied current, the surfaces in each column of Figure 10, except the column with the feed rate of 0.1 mm/s, become smooth, which can be regarded as the result of the polishing film thickening. In each row of Figure 10, the surfaces become smooth along with the increase of the feed rate. This is because, when the current density near the droplet trail is much lower than the center area, a lower feed rate means a longer time of exposure to the low current density and this contributes to the thinning of the supersaturated film. Another noticeable problem in Figure 10 is that, although Figure 10l appears more smooth compared to other surfaces, its roughness is higher than that in Figure 10k. This may be explained by the time resolved model of the surface structure during ECM proposed by Rosenkranz et al. [27]. The proposed theory is that for very long current pluses or even direct current, the crystallographic orientation of the grains influence highly the electrochemical dissolution process and the roughness depends on the dissolution speed of different grains. From this point of view, the exposure to the low current density is possibly playing a positive role, because it can thin down the uneven polishing film induced by a large current density and flatten the machined surface. This might explain why, with the applied current of 300 mA, the feed rate of 0.3 mm/s produces a lower roughness compared to the feed rate of 0.4 mm/s.

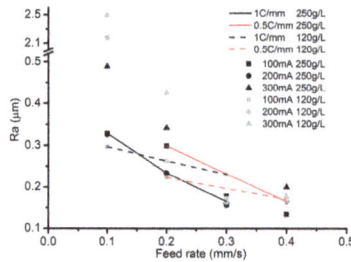

Figure 9. Roughness under different current values and concentrations.

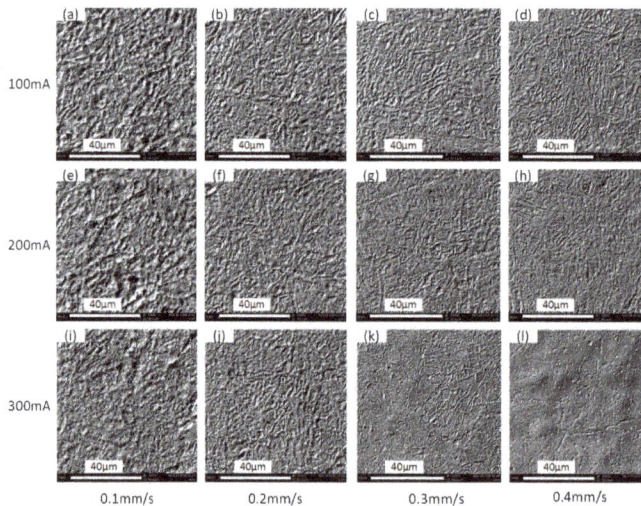

Figure 10. Microstructures of the channel bottom surface. (**a**) Current: 100 mA, feed rate: 0.1 mm/s; (**b**) current: 100 mA, feed rate: 0.2 mm/s; (**c**) current: 100 mA, feed rate: 0.3 mm/s; (**d**) current: 100 mA, feed rate: 0.4 mm/s; (**e**) current: 200 mA, feed rate: 0.1 mm/s; (**f**) current: 200 mA, feed rate: 0.2 mm/s; (**g**) current: 200 mA, feed rate: 0.3 mm/s; (**h**) current: 200 mA, feed rate: 0.4 mm/s; (**i**) current: 300 mA, feed rate: 0.1 mm/s; (**j**) current: 300 mA, feed rate: 0.2 mm/s; (**k**) current: 300 mA, feed rate: 0.3 mm/s; (**l**) current: 300 mA, feed rate: 0.4 mm/s.

3.3. Effects of Vacuum Gap (VG)

The effects of VG have been also evaluated under a uniform air pressure. VG sizes of 200 μm, 300 μm and 400 μm have been, respectively, selected in the experiments to investigate its influence on the machining performance. The applied current setting was 200 mA and the IEG was set to 50 μm. In Figure 11, it can be noticed that, under the same feed rate, the trail of the droplet above the workpiece top surface becomes longer as the VG increases.

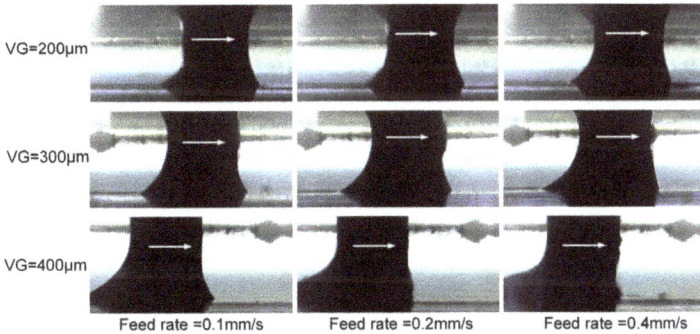

Figure 11. Droplet morphologies under different vacuum gap (VG).

Figure 12 is the plot of the current signal under the VG of 400 μm. Compared to the situation with the feed rate of 0.1 mm/s, violent current fluctuations (~40 mA) have been detected around 200 mA in the three larger feed rates. This is a sign of instability in the flow field, which can also be confirmed by their corresponding droplet morphologies in Figure 11.

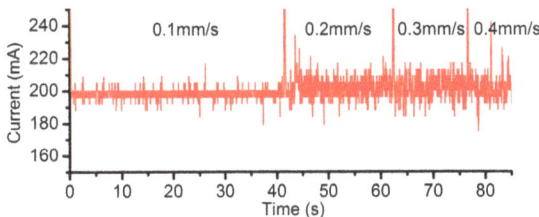

Figure 12. Current signal when setting 200 mA as the target (VG = 400 μm).

Figure 13 portrays the cross-sectional profiles of the channels machined under different VGs and different feed rates. It can be noticed that when the VG is set to 400 μm, the cross-sectional profiles become irregular, apart from the case with a feed rate of 0.1 mm/s. Therefore, it can be assumed that there is a correlation between the fluctuation of current signals (Figure 12) and the profiles of the machined channels. Figure 14 exhibits clearly the morphology difference of the channels machined with feed rates of 0.1 mm/s and 0.2 mm/s. It can be seen that the channel with 0.2 mm/s has a very rough surface. Therefore, a VG of 400 μm can lead to an unstable machining performance, except for the cases with very small feed rate. It is noticeable that when the VG is set to 200 μm and 300 μm, there is no significant deviation between them, although some deviations exist in the morphologies of the moving droplets, as shown in the microscopic graphs (Figure 11).

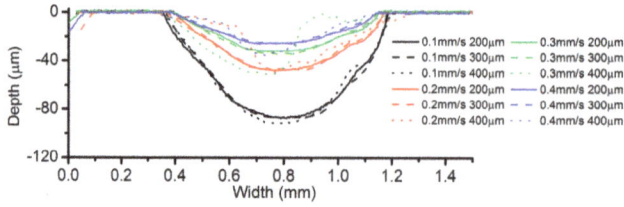

Figure 13. Cross-sectional profiles with different VGs (current = 200 mA).

Figure 14. Channel morphologies under the VG of 400 μm at: (**a**) 0.1 mm/s; and (**b**) 0.2 mm/s.

Table 2 demonstrates the current efficiency when varying the VG. There is no definite discipline found on whether the VG affects the current efficiency.

Table 2. Current efficiency.

VG (μm)	Feed Rate (mm/s)	ECPL (C/mm)	S_{exp} (mm^2)	S_{the} (mm^2)	η_1	η_2
200	0.1	2	0.0498	0.0466	106.9%	87.8%
200	0.2	1	0.0246	0.0233	105.6%	86.7%
200	0.3	0.667	0.0164	0.0155	105.8%	86.9%
200	0.4	0.5	0.0127	0.01165	109.0%	89.5%
300	0.1	2	0.0499	0.0466	107.1%	87.9%
300	0.2	1	0.0246	0.0233	105.6%	86.7%
300	0.3	0.667	0.0169	0.0155	109.0%	89.5%
300	0.4	0.5	0.0122	0.01165	104.7%	85.9%
400	0.1	2	0.0499	0.0466	107.1%	87.9%

The channel width under different VGs in Figure 15 indicates that a larger VG can lead to a larger channel width, especially in the case with low electrolyte concentration. Figure 16 reveals the roughness of the channel bottom surface under different VGs. With an electrolyte concentration of 250 g/L, a smaller VG (200 μm) results in a lower roughness. As for the electrolyte concentration of 120 g/L, the roughness trend shows an adverse effect.

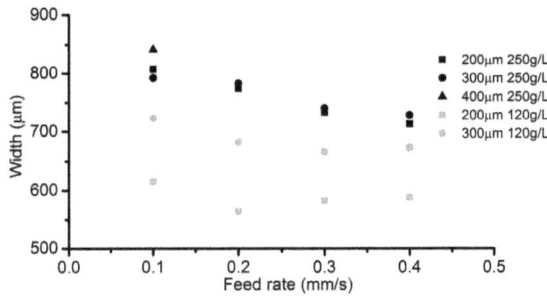

Figure 15. Channel width under different VGs.

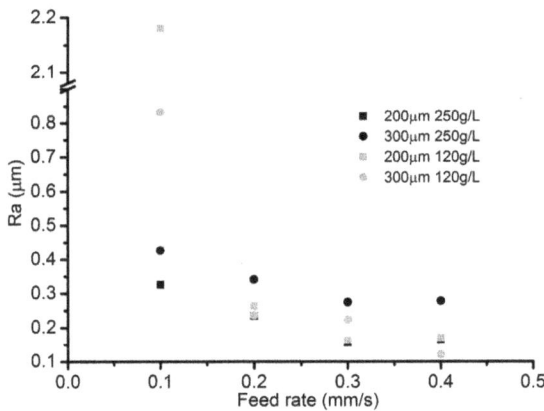

Figure 16. Roughness under different VGs.

Figure 17 shows the surface microstructures under different VGs while the same electrolyte concentration is used (i.e., 250 g/L). Obvious heterogeneous microstructures distribute on the same surface when the VG is set to 300 μm and 400 μm, which spoils surface conformance and roughness. One possible reason for this phenomenon is that a larger VG could lead to a slower transport process and an increase in the concentration of the products, which further influences the formation of microstructures. From this point of view, it is better to apply as small as possible a VG to accelerate the update of the electrolyte under the condition of stabilizing the electrolyte droplet. Since a feed rate larger than 0.2 mm/s under the VG of 400 μm will result in irregular channel morphologies (as depicted in Figure 14), those corresponding surface microstructures are demonstrated in a larger view in Figure 17 to show the uneven surface.

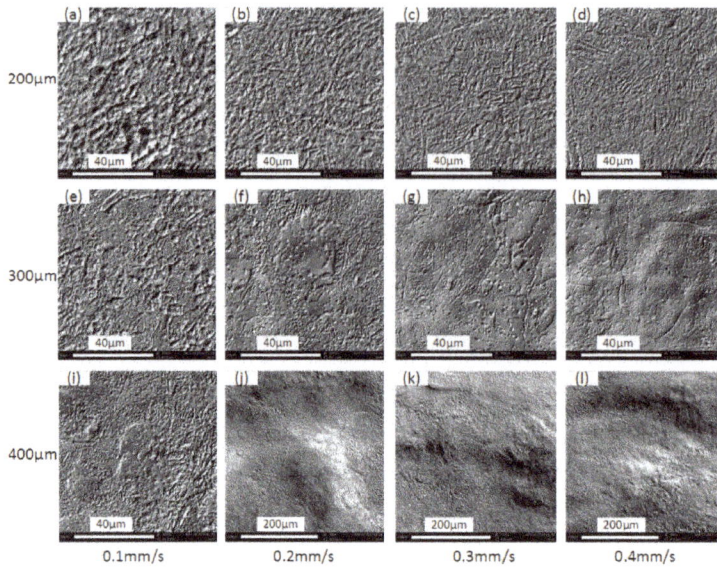

Figure 17. Surface microstructures under the current of 200 mA. (**a**) VG: 200 μm, feed rate: 0.1 mm/s; (**b**) VG: 200 μm, feed rate: 0.2 mm/s; (**c**) VG: 200 μm, feed rate: 0.3 mm/s; (**d**) VG: 200 μm, feed rate: 0.4 mm/s; (**e**) VG: 300 μm, feed rate: 0.1 mm/s; (**f**) VG: 300 μm, feed rate: 0.2 mm/s; (**g**) VG: 300 μm, feed rate: 0.3 mm/s; (**h**) VG: 300 μm, feed rate: 0.4 mm/s; (**i**) VG: 400 μm, feed rate: 0.1 mm/s; (**j**) VG: 400 μm, feed rate: 0.2 mm/s; (**k**) VG: 400 μm, feed rate: 0.3 mm/s; (**l**) VG: 400 μm, feed rate: 0.4 mm/s.

3.4. Effects of Inter-Electrode Gap (IEG)

During the preliminary experiments, it has been identified that only an IEG below 100 μm was capable of stabilizing the electrolyte droplet. Thus, IEGs of 50 μm and 80 μm have been selected for further investigation. Figure 18 illustrates the cross-sectional profiles under different IEGs. It appears that the profile of the channels machined with the same feed rate has very small deviations in its maximum depth. Figure 19 demonstrates the variation of the roughness and width of the machined channels. The channels machined with a VG of 80 μm are wider than those with a VG of 50 μm, no matter what the feed rate is. This phenomenon can be explained by the fact that the applied voltage will be automatically elevated to maintain the specified current value when the IEG increases, which will in turn reduce the contact angle of the droplet and widen the contact area because of the electrowetting. As for the roughness, its relationship with the IEG still needs to be studied.

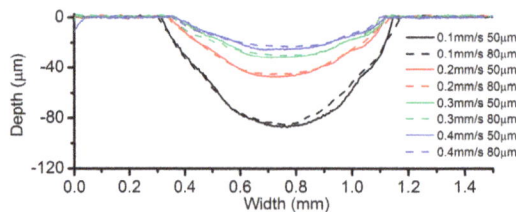

Figure 18. Cross-sectional profiles under different inter-electrode gaps (IEGs) (concentration: 250 g/L, current: 200 mA, VG: 200 μm).

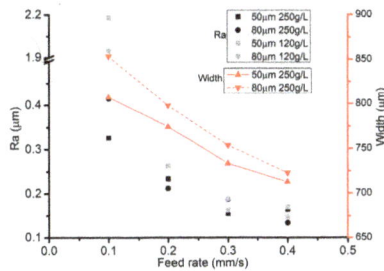

Figure 19. Channel roughness and width under different IEGs.

Table 3 shows the change of current efficiency with IEG. It can be noticed that the current efficiency in all cases with the IEG of 50 μm is larger than its counterpart at 80 μm.

Table 3. Current efficiency.

IEG (μm)	Feed Rate (mm/s)	ECPL (C/mm)	S_{exp} (mm^2)	S_{the} (mm^2)	η_1	η_2
50	0.1	2	0.0498	0.0466	106.9%	87.8%
80	0.1	2	0.0477	0.0466	102.4%	84.1%
50	0.2	1	0.0246	0.0233	105.6%	86.7%
80	0.2	1	0.0245	0.0233	105.6%	86.7%
50	0.3	0.667	0.0164	0.0155	105.8%	86.9%
80	0.3	0.667	0.0161	0.0155	103.9%	85.3%
50	0.4	0.5	0.0127	0.01165	109.0%	89.5%
80	0.4	0.5	0.0120	0.01165	103.0%	84.6%

3.5. Effects of Electrolyte Concentration

The effects of the electrolyte concentration on the machining performance have already been partially illustrated in the sections above. Additional experiments have been conducted to investigate the surface microstructures derived under different electrolyte concentrations. The cross-sectional profiles in Figure 20 indicate that the electrolyte concentration has barely any effect on the maximum depth of machined channels. Figure 21 describes the surface microstructures obtained under different electrolyte concentrations. When the current is set to 200 mA, there is not much difference in the surface structures under the same feed rate. While with a current of 300 mA, there exist clear differences between images in Figure 21k,o as well as the images in Figure 21l,p. This discrepancy implies that the polishing film thickness depends heavily on the electrolyte concentration under higher current densities. This also corresponds to the case in Figure 9, in which a lower roughness is obtained under the electrolyte concentration of 120 g/L in comparison to the electrolyte concentration of 250 g/L. In summary, a lower electrolyte concentration is beneficial towards a smooth bottom surface because an uneven, thick and supersaturated layer formed under a high concentration electrolyte can deteriorate the roughness.

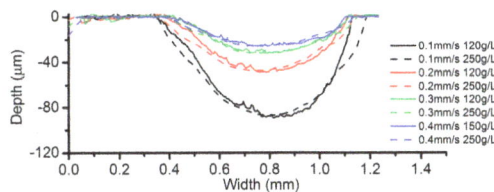

Figure 20. Cross-sectional profiles (current = 200 mA, VG = 200 μm).

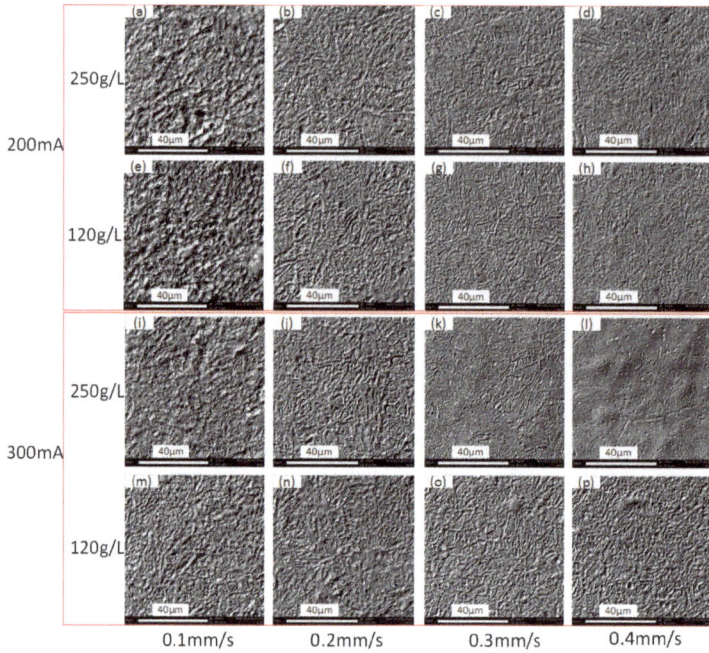

Figure 21. Microstructures of the channel bottom surface. (**a**) Current: 200 mA, electrolyte concentration: 250 g/L, feed rate: 0.1 mm/s; (**b**) current: 200 mA, electrolyte concentration: 250 g/L, feed rate: 0.2 mm/s; (**c**) current: 200 mA, electrolyte concentration: 250 g/L, feed rate: 0.3 mm/s; (**d**) current: 200 mA, electrolyte concentration: 250 g/L, feed rate: 0.4 mm/s; (**e**) current: 200 mA, electrolyte concentration: 120 g/L, feed rate: 0.1 mm/s; (**f**) current: 200 mA, electrolyte concentration: 120 g/L, feed rate: 0.2 mm/s; (**g**) current: 200 mA, electrolyte concentration: 120 g/L, feed rate: 0.3 mm/s; (**h**) current: 200 mA, electrolyte concentration: 120 g/L, feed rate: 0.4 mm/s; (**i**) current: 300 mA, electrolyte concentration: 250 g/L, feed rate: 0.1 mm/s; (**j**) current: 300 mA, electrolyte concentration: 250 g/L, feed rate: 0.2 mm/s; (**k**) current: 300 mA, electrolyte concentration: 250 g/L, feed rate: 0.3 mm/s; (**l**) current: 300 mA, electrolyte concentration: 250 g/L, feed rate: 0.4 mm/s; (**m**) current: 300 mA, electrolyte concentration: 120 g/L, feed rate: 0.1 mm/s; (**n**) current: 300 mA, electrolyte concentration: 120 g/L, feed rate: 0.2 mm/s; (**o**) current: 300 mA, electrolyte concentration: 120 g/L, feed rate: 0.3 mm/s; (**p**) current: 300 mA, electrolyte concentration: 120 g/L, feed rate: 0.4 mm/s.

The calculated current efficiency in Table 4 shows that higher concentration contributes to a higher current efficiency compared with its lower-concentration counterpart.

Table 4. Current efficiency.

Concentration (g/L)	Feed Rate (mm/s)	ECPL (C/mm)	S_{exp} (mm^2)	S_{the} (mm^2)	η_1	η_2
120	0.1	2	0.0473	0.0466	101.5%	83.3%
250	0.1	2	0.0498	0.0466	106.9%	87.8%
120	0.2	1	0.0242	0.0233	103.9%	85.3%
250	0.2	1	0.0246	0.0233	105.6%	86.7%
120	0.3	0.667	0.0152	0.0155	98.1%	80.5%
250	0.3	0.667	0.0164	0.0155	105.8%	86.7%
120	0.4	0.5	0.0116	0.01165	99.6%	81.8%
250	0.4	0.5	0.0127	0.01165	109.0%	89.5%

3.6. Multiple Processing

Similar to using jet-ECM to generate complex surfaces by superimposed multi-dimensional motion [28] and to derive specified waviness through parameters adjustment [29], it is also possible to fabricate a pocket with the SMEFC by overlapping multiple process trajectories. In the experiments, the VG was set to 200 μm and the IEG was set to 50 μm as default unless otherwise specified. An overlapping of seven paths was adopted in a zigzag model as shown in Figure 22. The current settings and ECPL utilized in the experiments are described in Table 5. The feed rates were determined corresponding to the setting of ECPL. Three electrode offset distances (EOD), i.e., 200 μm, 300 μm and 400 μm, have been chosen for the experiments.

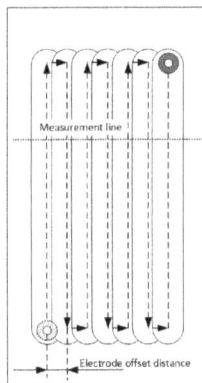

Figure 22. Electrode path.

Table 5. Feed rate (mm/s) used in the experiments.

ECPL (C/mm) Current (mA)	1	2/3	1/2	1/3	1/4
200	0.2	0.3	0.4	0.6	0.8
300	0.3	0.45	0.6	0.9	1.2
400	0.4	0.6	0.8	1.2	1.6

Figure 23 reveals the cross-sectional profiles of the cavities machined with different EODs with a current setting of 400 mA. It can be seen that the cavity depth increases as the EOD decreases, under the same feed rate. When the EOD is set to 400 μm (Figure 23c), there exist obvious periodic peaks on the bottom surface and the height of the peaks decrease along with the increase of the feed rate. Such kind of peaks cannot be observed in the profiles under the other two EODs. As shown in Figure 23a, when the feed rate is set at 1.6 mm/s, the bottom surface is relatively flat. However, in the case with four other feed rates, irregular spikes and valleys appear on the bottom surface. A possible reason for this phenomenon is that, with the same level of current, a small EOD will result in a relatively deep channel, which will in turn affect the flow field distribution of the neighboring electrolyte droplet. These effects tend to induce the electrolyte leakage and consequently deteriorate the bottom surface. A similar situation exists in Figure 24, which shows the cross-sectional profiles under the current of 300 mA and the EOD of 200 μm. When the EOD is set to 300 μm (Figure 23b), the bottom surface appears rather smooth except in the case feed rate of 0.4 mm/s.

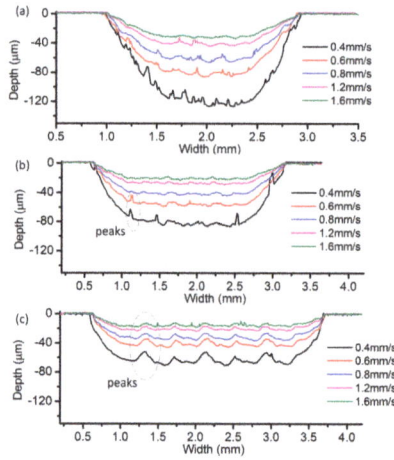

Figure 23. Cross-sectional profile with different electrode offset distances (EODs): (**a**) 200 μm; (**b**) 300 μm; and (**c**) 400 μm.

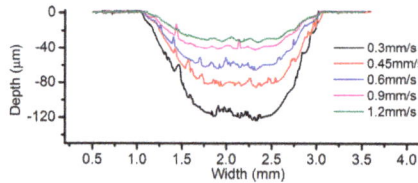

Figure 24. Cross-sectional profile with the current of 300 mA and EOD of 200 μm.

The volume removal rate (VRR) under different EODs has also been calculated and drawn in Figure 25. It can be found that the MRR with an EOD of 300 μm remains at stable values around 0.0257 mm^3/s, even if the ECPL changes from 1 C/mm to 1/4 C/mm. Considering both the flatness and the VRR stability, an EOD of 300 μm is a suitable value to be applied.

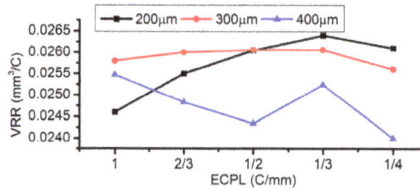

Figure 25. Volume removal rate (VRR) under different EODs.

The effect of the applied current has also been experimentally investigated, under the condition of the same ECPL. Figure 26 shows the comparison of the cross-sectional profiles under different current. It can be noticed that a better surface quality can be achieved under the current setting of 400 mA (Figure 26b). The case of 300 mA shows a rather rough profile and the corresponding current signal (Figure 27) fluctuates violently compared with other feed rates, which means the power supply needs more time to adjust the parameters to adapt the unstable flow field. When the EOD is set to 200 μm, rough cross-sectional profiles are formed no matter what the current setting is. To clarify this phenomenon, Figure 28 shows the bottom surface of the cavities machined under different EODs and the same ECPL of 1 C/mm. There are obvious residues on the surface of Figure 28a–c. Even with an EOD of 300 μm, residues still can be observed (in Figure 28d,e), but the cavity under the current setting of 400 mA shows a relatively smooth surface without any material residues as depicted in Figure 28f.

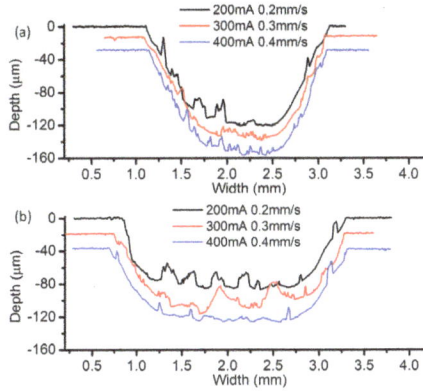

Figure 26. Cross-sectional profiles with different EODs: (**a**) 200 μm; (**b**) 300 μm.

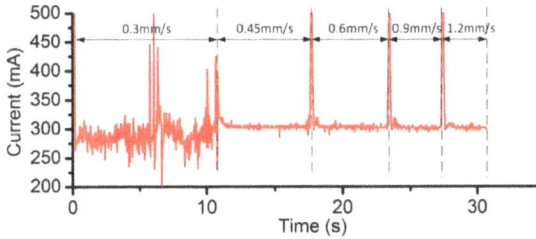

Figure 27. Current signal under 300 mA.

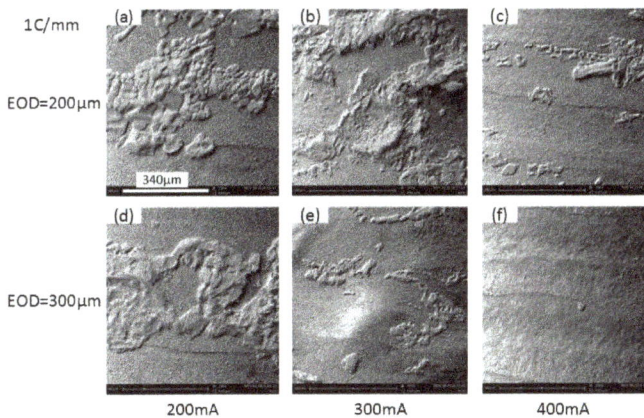

Figure 28. SEM pictures of the machined cavities. (**a**) EODs: 200 μm, current: 200 mA; (**b**) EODs: 200 μm, current: 300 mA; (**c**) EODs: 200 μm, current: 400 mA; (**d**) EODs: 300 μm, current: 200 mA; (**e**) EODs: 300 μm, current: 300 mA; (**f**) EODs: 300 μm, current: 400 mA.

In order to determine the residues composition, energy dispersive spectroscopy (EDS) analysis was conducted on a JXA-8530F (Jeol, Peabody, MA, UAS). Figure 29 shows the element spectrum and obvious peaks in Na element and O element were found in Region 001. It can be concluded that the residues on the machined surface is crystallization of $NaNO_3$ salt. Figure 30 shows the schematic shape of the electrolyte droplet along the measurement line during the electrochemical dissolution. When EOD is 200 μm, there are more electrolytes in the cavity than the case with the EOD of 300 μm. It is more possible to generate electrolyte leakage when EOD is smaller. The reason why larger current induce less residues may be that, at the same level of ECPL, larger current means larger feed rate, which reduces the time of the electrolyte exposure to the fast airflow. This reduction inhibits the electrolyte crystallization possibility.

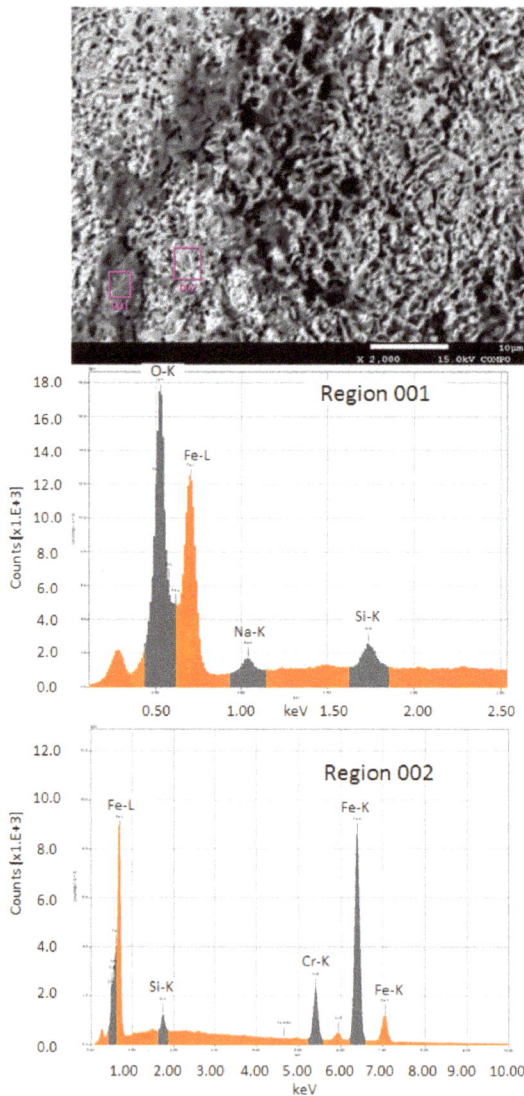

Figure 29. Energy-dispersive spectroscopy (EDS) analysis of the residues.

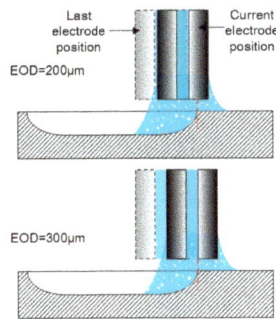

Figure 30. Schematic of different EODs.

Figure 31 shows the maximum cavity depth (d_{max}) obtained under different current settings. When the ECPL is kept the same, the d_{max} always remains the same, although a higher current setting value will result in several microns larger than the d_{max} derived by the lower current values.

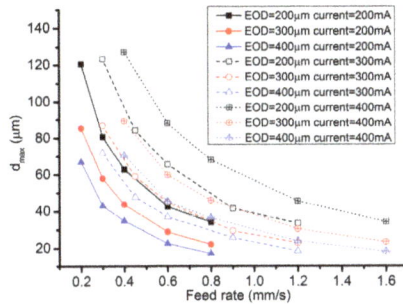

Figure 31. Maximum depth under different current values.

Table 6 shows the quantitative fitting between the ECPL (q) and the d_{max} at different combinations of the current and the EOD. It appears that the cavity depth has nearly a linear relationship with the ECPL variation, since the exponent of q is close to 1. With the increase of the current setting values, the maximum depth always grows in a small range.

Table 6. d_{max} (µm) fitting equations.

Current (mA) \ EOD (µm)	200	300	400
200	$120 \times q^{0.9365}$	$85.49 \times q^{0.9714}$	$66.5 \times q^{0.9821}$
300	$123.7 \times q^{0.9516}$	$87.17 \times q^{0.9657}$	$71.72 \times q^{0.9607}$
400	$127.8 \times q^{0.9326}$	$89.41 \times q^{0.9702}$	$70.15 \times q^{0.9807}$

4. Conclusions

1. An integrable scanning micro electrochemical flow cell (SMEFC) unit has been developed and utilized to fabricate mesoscale channels. The SMEFC can confine the electrolyte droplet just in the area of around 0.5 mm^2 without leakage to the other non-processing region.
2. The roughness of the original workpiece surface affects the shape of the moving electrolyte droplet. Smooth surfaces tend to induce longer trials of the electrolyte droplet above the top of the workpiece surface.

3. Among all the geometrical parameters of the SMEFC configuration, the vacuum gap significantly affects the shape of the moving electrolyte droplet. A smaller vacuum gap tends to provide a better control of the electrolyte droplet, which contributes to the consistency of the surface microstructures. At the same level of the electric consumption per unit length, the channel depth maintain the same on the whole, but suitable combinations of the current densities and feed rates can generate better surface quality and roughness. The concentration of the electrolyte influences the formation of the supersaturated layer and further affects the roughness. A higher current density, a smaller inter-electrode gap and a higher electrolyte concentration improve the current efficiency.

4. As for the electrochemical milling by the SMEFC, the cavities machined with different electrode offset distances and different electric consumption per unit length have been compared. The electrode offset distance plays a significant role on the milling performance. Through the comparison of the cross-sectional profiles and SEM pictures, the electrode offset distance of 300 μm together with the current of 400 mA is a better combination for obtaining a relatively smooth bottom surface and a stable material removal rate. The removal depth can be adjusted by tuning the feed rate. A larger electrode offset distance and a higher current contribute to decrease of the residues of the electrolyte crystallization.

5. In future research, a linear power supply can replace the switching power supply for a better current holding capability. The electrode with the outer diameter of 0.3 mm will also be utilized to obtain smaller features. Multiple vision units will be installed at different angles to monitor and analyze the moving electrolyte droplet.

Acknowledgments: The authors would like to thank the China Scholarship Council (CSC) for providing a scholarship. Part of this research has been financed by the EC project Hi-Micro (Grant No. 314055) and MicroMAN (Grant No. 674801). The authors also would like to thank the technicians and Yansong Guo at the Mechanical Engineering Department of KU Leuven, for their assistance on component fabrication, and Shuigen Huang for the EDS analysis.

Author Contributions: Under the supervision of Dominiek Reynaerts and Jun Qian, Cheng Guo conducted the main research work, including hardware and software development of the experimental setup, design of the experiments, analysis of the results, and writing the paper. Dominiek Reynaerts and Jun Qian have facilitated the experiments, provided daily guidance and suggestions for this research work, and revised the paper.

Conflicts of Interest: The authors declare no conflict of interest. The founding sponsors had no role in the design of the study; in the collection, analyses, or interpretation of data; in the writing of the manuscript, and in the decision to publish the results.

References

1. Lee, S.-J.; Lee, C.-Y.; Yang, K.-T.; Kuan, F.-H.; Lai, P.-H. Simulation and fabrication of micro-scaled flow channels for metallic bipolar plates by the electrochemical micro-machining process. *J. Power Sources* **2008**, *185*, 1115–1121. [CrossRef]

2. Hung, J.-C.; Chang, C.-H.; Chiu, K.-C.; Lee, S.-J. Simulation-based fabrication of micro-helical grooves in a hydrodynamic thrust bearing by using ecmm. *Int. J. Adv. Manuf. Technol.* **2013**, *64*, 813–820. [CrossRef]

3. Liu, G.X.; Zhang, Y.J.; Jiang, S.Z.; Liu, J.W.; Gyimah, G.K.; Luo, H.P. Investigation of pulse electrochemical sawing machining of micro-inner annular groove on metallic tube. *Int. J. Mach. Tools Manuf.* **2016**, *102*, 22–34. [CrossRef]

4. Ryu, S.H. Eco-friendly ecm in citric acid electrolyte with microwire and microfoil electrodes. *Int. J. Precis. Eng. Manuf.* **2015**, *16*, 233–239. [CrossRef]

5. Jia, L.; Xiaochen, J.; Di, Z. Electrochemical machining of multiple slots with low-frequency tool vibrations. *Procedia CIRP* **2016**, *42*, 799–803. [CrossRef]

6. Liu, G.; Zhang, Y.; Deng, Y.; Wei, H.; Zhou, C.; Liu, J.; Luo, H. The tool design and experiments on pulse electrochemical machining of micro channel arrays on metallic bipolar plate using multifunctional cathode. *Int. J. Adv. Manuf. Technol.* **2017**, *89*, 407–416. [CrossRef]

7. Natsu, W.; Ikeda, T.; Kunieda, M. Generating complicated surface with electrolyte jet machining. *Precis. Eng.* **2007**, *31*, 33–39. [CrossRef]

8. Hackert, M.; Meichsner, G.; Schubert, A. Generating micro geometries with air assisted jet electrochemical machining. In Proceedings of the Euspen 10th Anniversary International Conference, Zurich, Switzerland, 18–22 May 2008; pp. 420–424.

9. Kai, S.; Sai, H.; Kunieda, M.; Izumi, H. Study on electrolyte jet cutting. *Procedia CIRP* **2012**, *1*, 627–632. [CrossRef]

10. Kunieda, M.; Mizugai, K.; Watanabe, S.; Shibuya, N.; Iwamoto, N. Electrochemical micromachining using flat electrolyte jet. *CIRP Ann. Manuf. Technol.* **2011**, *60*, 251–254. [CrossRef]

11. Hackert-Oschätzchen, M.; Meichsner, G.; Zinecker, M.; Martin, A.; Schubert, A. Micro machining with continuous electrolytic free jet. *Precis. Eng.* **2012**, *36*, 612–619. [CrossRef]

12. Ghoshal, B.; Bhattacharyya, B. Investigation on profile of microchannel generated by electrochemical micromachining. *J. Mater. Process. Technol.* **2015**, *222*, 410–421. [CrossRef]

13. Ghoshal, B.; Bhattacharyya, B. Electrochemical micromachining of microchannel using optimum scan feed rate. *J. Manuf. Process.* **2016**, *23*, 258–268. [CrossRef]

14. Kim, B.H.; Ryu, S.H.; Choi, D.K.; Chu, C.N. Micro electrochemical milling. *J. Micromech. Microeng.* **2005**, *15*, 124. [CrossRef]

15. Zhang, Z.; Zhu, D.; Qu, N.; Wang, M. Theoretical and experimental investigation on electrochemical micromachining. *Microsyst. Technol.* **2007**, *13*, 607–612. [CrossRef]

16. Shin, H.S.; Kim, B.H.; Chu, C.N. Analysis of the side gap resulting from micro electrochemical machining with a tungsten wire and ultrashort voltage pulses. *J. Micromech. Microeng.* **2008**, *18*, 075009. [CrossRef]

17. Wang, S.; Zhu, D.; Zeng, Y.; Liu, Y. Micro wire electrode electrochemical cutting with low frequency and small amplitude tool vibration. *Int. J. Adv. Manuf. Technol.* **2011**, *53*, 535–544. [CrossRef]

18. Liu, Z.; Nouraei, H.; Spelt, J.K.; Papini, M. Electrochemical slurry jet micro-machining of tungsten carbide with a sodium chloride solution. *Precis. Eng.* **2015**, *40*, 189–198. [CrossRef]

19. Yuan, L.; Xu, J.; Zhao, J.; Zhang, H. Research on hybrid process of laser drilling with jet electrochemical machining. *J. Manuf. Sci. Eng.* **2012**, *134*, 064502. [CrossRef]

20. Zhang, Z.; Cai, M.; Feng, Q.; Zeng, Y. Comparison of different laser-assisted electrochemical methods based on surface morphology characteristics. *Int. J. Adv. Manuf. Technol.* **2014**, *71*, 565–571. [CrossRef]

21. Sakairi, M.; Sato, F.; Gotou, Y.; Fushimi, K.; Kikuchi, T.; Takahashi, H. Development of a novel microstructure fabrication method with co-axial dual capillary solution flow type droplet cells and electrochemical deposition. *Electrochim. Acta* **2008**, *54*, 616–622. [CrossRef]

22. Drensler, S.; Milenkovic, S.; Hassel, A.W. Microvials with tungsten nanowire arrays. *J. Solid State Electrochem.* **2014**, *18*, 2955–2961. [CrossRef]

23. Hu, J.-F.; Kuo, C.-L. Study on Micro Electrochemical Machining Using Coaxial for Gushing and Sucking Method. Master's Thesis, National Yunlin University of Science & Technology, Yunlin County, Taiwan, 2007.

24. Kuo, K.-Y.; Wu, K.-L.; Yang, C.-K.; Yan, B.-H. Wire electrochemical discharge machining (WECDM) of quartz glass with titrated electrolyte flow. *Int. J. Mach. Tools Manuf.* **2013**, *72*, 50–57. [CrossRef]

25. Leroy, P.; Lassin, A.; Azaroual, M.; André, L. Predicting the surface tension of aqueous 1:1 electrolyte solutions at high salinity. *Geochim. Cosmochim. Acta* **2010**, *74*, 5427–5442. [CrossRef]

26. Deconinck, D.; Hoogsteen, W.; Deconinck, J. A temperature dependent multi-ion model for time accurate numerical simulation of the electrochemical machining process. Part III: Experimental validation. *Electrochim. Acta* **2013**, *103*, 161–173. [CrossRef]

27. Rosenkranz, C.; Lohrengel, M.; Schultze, J. The surface structure during pulsed ecm of iron in $NaNO_3$. *Electrochim. Acta* **2005**, *50*, 2009–2016. [CrossRef]

28. Schubert, A.; Hackert-Oschätzchen, M.; Martin, A.; Winkler, S.; Kuhn, D.; Meichsner, G.; Zeidler, H.; Edelmann, J. Generation of complex surfaces by superimposed multi-dimensional motion in electrochemical machining. *Procedia CIRP* **2016**, *42*, 384–389. [CrossRef]

29. Martin, A.; Eckart, C.; Lehnert, N.; Hackert-Oschätzchen, M.; Schubert, A. Generation of defined surface waviness on tungsten carbide by jet electrochemical machining with pulsed current. *Procedia CIRP* **2016**, *45*, 231–234. [CrossRef]

micromachines

MDPI

Article

On the Application of Replica Molding Technology for the Indirect Measurement of Surface and Geometry of Micromilled Components

Federico Baruffi [1,*]**, Paolo Parenti** [2]**, Francesco Cacciatore** [2]**, Massimiliano Annoni** [2] **and Guido Tosello** [1]

[1] Department of Mechanical Engineering, Technical University of Denmark, Produktionstorvet, Building 427A, 2800 Kgs. Lyngby, Denmark; guto@mek.dtu.dk
[2] Department of Mechanical Engineering, Politecnico di Milano, via la Masa 1, 20100 Milan, Italy; paolo.parenti@polimi.it (P.P.); francesco.cacciatore@polimi.it (F.C.); massimiliano.annoni@polimi.it (M.A.)
* Correspondence: febaru@mek.dtu.dk; Tel.: +45-4525-4822

Received: 24 May 2017; Accepted: 18 June 2017; Published: 21 June 2017

Abstract: The evaluation of micromilled parts quality requires detailed assessments of both geometry and surface topography. However, in many cases, the reduced accessibility caused by the complex geometry of the part makes it impossible to perform direct measurements. This problem can be solved by adopting the replica molding technology. The method consists of obtaining a replica of the feature that is inaccessible for standard measurement devices and performing its indirect measurement. This paper examines the performance of a commercial replication media applied to the indirect measurement of micromilled components. Two specifically designed micromilled benchmark samples were used to assess the accuracy in replicating both surface texture and geometry. A 3D confocal microscope and a focus variation instrument were employed and the associated uncertainties were evaluated. The replication method proved to be suitable for characterizing micromilled surface texture even though an average overestimation in the nano-metric level of the *Sa* parameter was observed. On the other hand, the replicated geometry generally underestimated that of the master, often leading to a different measurement output considering the micrometric uncertainty.

Keywords: replica technology; roughness; micromilling; surface metrology; dimensional micro metrology

1. Introduction

In recent decades, the demand for micro components has steadily increased in many engineering sectors such as biotechnology, avionics, medicine, automotive, etc. [1,2]. With the aim of meeting the new challenging requirements in terms of precision and accuracy, the world of manufacturing reacted either by developing brand new technologies or by downscaling well-established ones. Micromilling belongs to the second class of processes, being the miniaturized adaptation of conventional milling technology. The ability to produce full three-dimensional micro components with relatively high material removal rate makes this process suitable for producing molds and inserts employed in replication techniques such as micro injection molding [3]. When evaluating the quality of a machined mold, two main characteristics must be addressed, namely the geometrical accuracy and the surface topography. In fact, mold dimensional features and surface topographies are directly transferred to replicated products [4]. Furthermore, the mold surface can directly affect the replication process since it influences, for instance, demolding forces in micro injection molding [5]. A comprehensive and detailed knowledge of micromilled mold surfaces and geometries is therefore necessary for optimizing the product performance.

When dealing with micromilling, numerous phenomena, negligible for conventional milling, become significant. One of them is related to the minimum chip thickness effect [6]. When machining with chip thickness values comparable to the tool edge radius, effects such as plowing and elastic recovery become dominant, causing a decrease of dimensional accuracy and an increase of surface roughness, in particular for soft-state metallic alloys [7].

At present, many dimensional and topographical measurement solutions can be applied to micro components [8]. Among all of them, optical instruments are strongly emerging [9] because of their peculiar advantages. Firstly, their contact-less and non-destructive nature makes them suitable for measuring soft and very small components: even a small force would in fact invalidate the measurement or damage the sample. Furthermore, most optical measuring techniques allow an areal surface characterization, which has more statistical significance in comparison to a profile one [10]. Finally, they are potentially faster than typical contact instruments such as coordinate-measuring machines (CMMs) and stylus profilometers. However, certain features, for instance micro holes or micro cavities, are inaccessible for optical lenses, since an insufficient amount of light is reflected back to the measuring objective. In other cases, samples may not physically fit under the microscope. Inaccessible features are typical in micro molds for injection molding applications, where high aspect ratio features must often be replicated [11–13]. In such cases, new solutions must be employed to perform dimensional and surface measurements without adopting destructive inspections. Among them, replica molding technology is the most promising one. This method is based on the replication of inaccessible features by means of casting-like soft polymeric media such as polydimethylsiloxane (PDMS) or two-component silicone polymers [4]. The common replication procedure simply consists in casting the replication media over the component to be replicated. After the curing is completed, the replica is removed and then measured, providing and indirect quantitative information of the actual geometry and surface. The use of such method is well-established in medical and dental fields [14,15] where specimens simply cannot be placed under a measuring instrument. It is also widely employed in non-destructive metallographic analysis of mechanical components [16,17]. For dimensional and surface metrology tasks, replication kits based on a two-component polymer are nowadays widely used [18,19]. Silicone based materials are usually employed, since they ensure almost no shrinkage over a wide range of curing temperature. When using these kits, the polymer and the curing agent are automatically mixed in a disposable nozzle. A dispensing gun is utilized for directing the casting of the replication media. Tosello et al. [20] successively employed this method to monitor the tool wear of a mold for Fresnel lenses production presenting micro structures with a height of 23 µm.

In recent decades, few research papers investigated the replication performance of fast replication media used in indirect metrology tasks. Madsen et al. [21] studied the shrinkage of a PDMS material utilized for measuring nano-metric geometries. The authors considered several mixing ratios and curing temperature levels and investigated the repeatability of the process through repeated replicas. The authors concluded that the utilized PDMS shrinks linearly between 1% and 3 %, leading to dimensions that are smaller than the ones of the replicated master. In a recent paper, Goodall et al. [22] analyzed the accuracy and precision of seven different silicone based replication media utilized for quantitative surface texture characterization. Two masters, one smooth and one rough lower jaw teeth, were replicated. Most of the areal surface texture parameters and fractal parameterization were employed to carry out comparisons. It was found that low viscosity media generally achieve higher accuracy and precision. However, the authors highlighted a need for standardization, since results from different impression media were not comparable. Finally, Gasparin et al. [23] investigated the surface texture replication performance of three replication media (one hard and two soft). Two calibrated roughness standards were used as masters. Measurements were performed with optical and tactile instruments. A replication degree up to 96% was reached, proving that the measured deviations fell inside the uncertainty range. Since the nominal *Ra* value of the measured standard was 500 nm, the authors' considerations cannot be directly extended to micromilled surfaces, which are usually smoother (in most applications, *Ra* ranges between 50 nm and 250 nm [24–27] on both flat and free-form parts).

Despite several approaches have been proposed in literature [28,29], the use of surface topography prediction models in industrial micromilling applications is still extremely limited. This is due to the lack of robustness that the models show with respect to the variability of the material characteristics, tool micro-geometry and tool wear. Therefore, the measurement of micromilled surface after machining represents the standard practice that machinists still adopt for verifying the cutting process results. The replication and the following indirect measurement are thus the preferable solution for characterizing the surface topography of inaccessible micromilled features.

At present, the literature still lacks a study focused on the performance verification of a replica procedure specifically applied to micromilled components.

The aim of this paper is to evaluate the performance of a commercial silicone replication media (RepliSet [19], Struers®, Ballerup, Denmark) in the indirect measurement of micromilled surfaces and geometries made from mold steel. In order to accomplish this, a quantitative comparison between micromilled masters and their replicas was carried out. In particular, the performance in replicating both surface topography and geometry was assessed separately using two different benchmark samples.

2. Materials and Methods

In order to assess the surface replication fidelity of the silicone media, two different samples were designed and then machined.

The first one was used to assess the replication performance related to indirect surface roughness measurements. Different benchmark surfaces were machined and then replicated. Micro-roughed and micro-finished surfaces were obtained by varying the radial depth of cut. Two materials were milled: AISI 440 hardened (AISI 440 H, hardness = 60 HRC) and AISI 440 annealed (AISI 440 A, hardness = 18 HRC). These two stainless steels represent a suitable choice for mold manufacturing and are expected to produce different surface topographies in relation to the specific material characteristics, as for instance hardness, grain size and specific cutting force. In particular, being AISI 440 A in a soft state, the risk of plowing is certainly higher than for the harder AISI 440 H [7].

The second sample was used to investigate the replication capability related to indirect geometrical measurements. Five micromilled pockets were produced on the same two aforementioned materials and successively replicated. The geometry of the pockets was varied in order to assess the replication fidelity of the silicone media at different penetration depths.

2.1. Micromilled Surfaces

The consideration of surface topography is the first step for the assessment of any replication method. In micromilling surface generation, several factors affect the final surface topography. In particular, milling process parameters, milling strategy and the geometry of the adopted tool all play an important role. In this study, three typologies of micromilled surface were produced on both the materials. A coated WC Round End Mill (Mitsubishi Materials Corporation, Tokyo, Japan) (2 flutes, diameter = 1 mm, Corner Radius = 0.1 mm and cutting edge radius = 6 μm) was used for the tests on an ultra-high precision KERN Evo micromilling machine. Table 1 shows the cutting parameters for the three machined surfaces.

Table 1. Cutting parameters for the three machined surfaces.

Cutting Parameter	S1	S2	S3
a_e/D_C	100% (full slot)	30% (overlapped)	30% (finished)
$a_p/\mu m$	50	50	50
$v_c/(m/min)$	100	100	100
$f_z/\mu m$	12.5	12.5	12.5

The first surface, designated S1, was machined in full slot (i.e., with radial depth of cut a_e equal to 100% of the mill diameter D_C). S2 and S3 were instead obtained by imposing a 30% radial depth of cut. S2 designates the overlapped surface area, generated by two subsequent tool passes. S3 is generated by the last mill pass. Figure 1 shows a scheme of the surface design. These specific surfaces represent typical conditions of a mold manufacturing process: S1 conditions are distinctive of roughing operations (e.g., during pocket milling), where full slot machining is employed to minimize the machining time. S2 and S3 are representative of finishing operation, where the final surface topography and appearance are generated using a lower value of the radial depth of cut. S1 surface roughness is therefore expected to be higher than for S2 and S3.

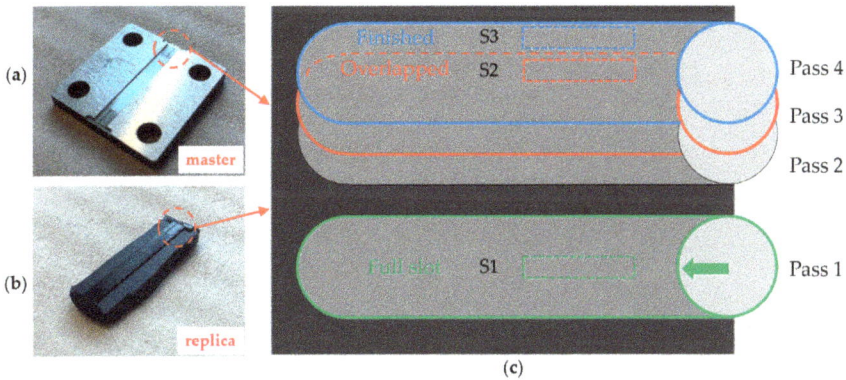

Figure 1. Micromilled surfaces and their replicas: (**a**) AISI 440 H sample. The two replicates of the machined surfaces are visible. (**b**) Silicone replica of the AISI 440 H sample. (**c**) Scheme of the micromilled surfaces. The four mill passes are presented in their machining sequence. The three measured surfaces S1, S2 and S3 are indicated with a dashed line. The green arrow shows the milling direction.

The other milling parameters were kept constant in the three cases. The feed per tooth f_z was set at the lowest limit for the selected mill. The axial depth of cut a_p was limited to 50 µm for all the tests to avoid the onset of chatter conditions which can lead to uncontrolled and defective surface generation. In order to incorporate the effect of the tool wear in the surface generation, the same tool was used for machining all the samples.

Two replicates for each surface type, designated as Sample 1 and Sample 2, were generated on the same metal workpiece (see Figure 1a). Six surfaces per material were therefore available for replication.

2.2. Micromilled Geometrical Features

In order to assess the performance of the replication media in terms of micromilled geometry replication, five micromilled pockets were produced on the two materials. Such features represent a generalization of two-dimensional channels and cavities that are often present in micro molds. For carrying out the machining, a WC flat end mill (Sandvik Coromant, Sandviken, Sweden) (2 flutes, diameter = 0.5 mm and cutting edge radius = 6 µm) was used on the same micromilling machine. The cutting parameters (see Table 2) were kept constant for all the machined pockets. Once again, the same tool was utilized for producing all the samples.

Table 2. Cutting parameters for the five machined pockets.

Cutting Parameter	Value
a_e/mm	200
a_p/μm	150
v_c/(m/min)	60
f_z/μm	10

Figure 2 shows the geometrical characteristics of the micromilled pockets. In particular, the five two-stepped square pockets were carried out with nominal constant width. The height of the two steps was instead varied in order to investigate the replication performance of the silicone when penetrating geometrical features with different depths. Moreover, the heights were set at a value that allows the micromilled pockets to be measured directly on the metal samples, thus making the comparison between original and replicated specimens feasible.

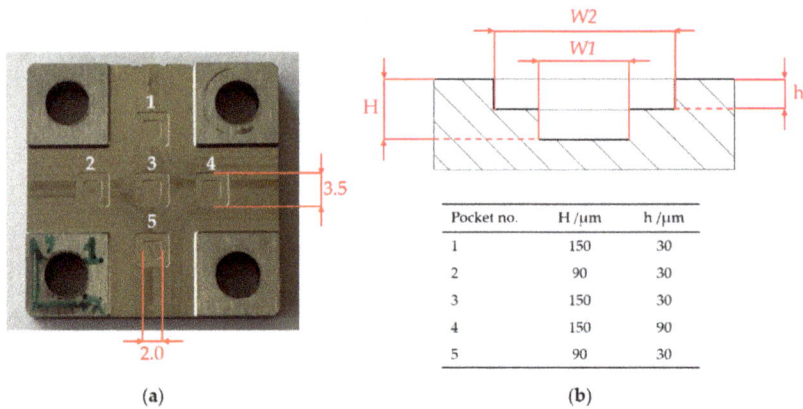

Pocket no.	H /μm	h /μm
1	150	30
2	90	30
3	150	30
4	150	90
5	90	30

(a)   (b)

Figure 2. (a) Micromilled pockets with nominal width dimensions; and (b) pocket depths and the two measurands *W1* and *W2*.

The width of the cavity was selected as reference for the replication assessment. Therefore, the dimensions *W1* and *W2* (see Figure 2) were measured on both metal master and replicated samples and then compared.

2.3. Replication Procedure

After the milling operation, the samples were cleaned from metal debris and dirty particles with ultrasonic cleaning and then blown with filtered air. The black two-component silicone rubber [19] was then poured on the steel masters by using the appropriate dispensing gun and following the supplier guidelines. The curing of the replication media was performed at room temperature. After complete solidification, the replicas were carefully removed from the master and prepared for the measurement, avoiding any additional contamination of the samples. In order to assess the repeatability of the replication, the procedure was repeated three times for each steel sample, resulting in three silicone replicas for each micromilled surface and pocket.

2.4. Surface Measurement Methodology and Uncertainty Evaluation Procedure

The surface topography measurements were carried out using a 3D confocal microscope (MarSurf CMW 100 from Mahr GmbH, Göttingen, Germany) with integrated white light interferometer employing a high-power 505 nm LED as light source. Both the measuring principles were employed in this study. Confocal microscopy is particularly suitable for 3D measurements of fine surfaces since it allows eliminating out-of-focus blurs by means of a selective pinhole [30]. The large numerical aperture of confocal microscopes objectives allows a very high maximum detectable slope (up to 75° [8]), which is particularly useful when dealing with the steep surface textures as those obtained by the micromilling process. Table 3 shows the main characteristics of the instrument.

Table 3. Confocal microscope characteristics.

Objective Magnification	100×
Numerical aperture	0.90
Working distance in mm	1.0
Field of view in µm	192 × 144
Optical lateral resolution in µm	0.28
Digital lateral resolution in µm	0.25
Vertical resolution in nm	1.0

With the aim of characterizing a significant portion of the micromilled surface texture, a rectangular area of 1.0 mm × 0.2 mm was acquired for each steel and silicone surface. In particular, a stitching operation (with 9 images) was automatically performed for acquiring the entire surface extension. Such operation can be successfully applied to surface topography analysis, since it only has a minor influence on the results [31]. The acquisition of a relatively large area also decreases the effect of potential relocations errors on the comparison between masters and replicas. The position of S1, S2 and S3 was defined with respect to a fixed planar reference position univocally defined on both steel and silicone samples. The identification of such a reference is particularly important for comparing milled surfaces, which present a continuous surface texture. The raw surface acquisitions were post-processed with a dedicated image metrology software (MountainsMap® [32], Digital Surf, Besançon, France). The images were initially flattened by means of a first order plane to ensure a correction for potential tilt of the samples. A peak-removal masking was also applied to remove a limited number of spikes that characterized the measured silicone surfaces. Due to their sharpness, these defects were considered as optical artifacts and therefore excluded from the quantitative texture analysis. Figure 3 shows the appearance of the three different micromilled surfaces. As expected, they present the typical characteristic of milled surfaces but with different patterns.

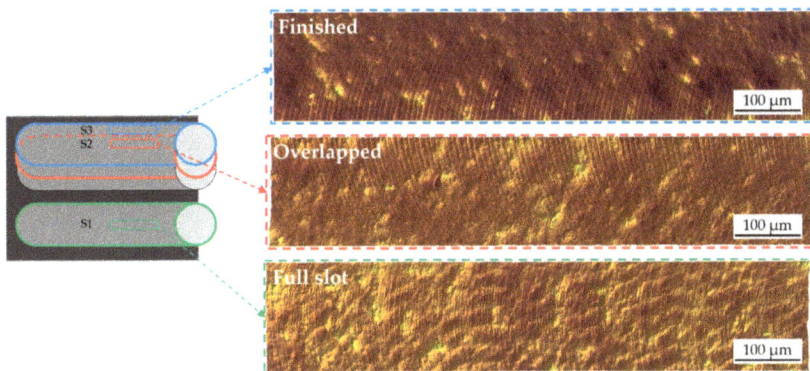

Figure 3. Surface texture appearance for the three acquired surfaces (AISI 440 H, Sample 2).

Finally, the performance verification of the replica technique was carried out by comparing masters and replicated surfaces using a synthetic surface parameter. In this case, the main objective was to assess the replication performance of the two-component silicone media with respect to the vertical profile development, since, given its nano-metric range, it is most critical to replicate. Therefore, the arithmetical mean height *Sa* was computed for each surface and then utilized as the parameter of comparison. This parameter, according to the ISO 25178-2 standard [33], gives indication about the average areal surface roughness of the surface, being the analogous of *Ra* that is used to describe profile measurements. *Sa* was chosen as indicator since it is only slightly influenced by local defects or optical artifacts, and therefore is suitable for a carrying out a general comparison between two surfaces. No cutoff filter was applied before calculating *Sa*. In this way, the performance replication throughout the entire spatial frequency domain was investigated.

To verify the quality of the measurements, an uncertainty evaluation was carried out. The measurement uncertainty *U* is a parameter associated with the results of a measurement that characterizes the dispersion of the values that could be reasonably be attributed to the measurand [34]. It is of paramount importance to determine this parameter and to include it in the evaluation of the replication capability, since, at the nano-scale, the variations due to replication process, instrument accuracy and measurement repeatability are in the same order of magnitude [4]. The uncertainty budget of the *Sa* measurements on the metal and silicone samples was estimated following ISO 15530-3 [35]. Although this method was conceived for CMM measurements, its principle can be successfully applied to optical measurements [23], allowing to avoid some of the complications that a full uncertainty estimation would introduce [36]. Such an approach is based on the substitution method, which allows estimating the instrument uncertainty by repeated measurements on a calibrated artifact sharing similar characteristic with the actual measurand. In this particular case, a calibrated roughness artifact (nominal value: *Ra* = 480 nm) was employed for this task. The measurement uncertainty related to surface roughness measurements when using the confocal microscope u_{CONF} was calculated as follows:

$$u_{CONF} = \sqrt{u^2_{cal} + u^2_p + u^2_{res,CONF}} \tag{1}$$

where u_{cal} is the standard calibration uncertainty of the roughness standard; u_p represents the standard uncertainty related to the measurement procedure and is calculated as standard deviation of fifteen repeated measurements on the calibrated standard; and $u_{res,CONF}$ is the resolution uncertainty related to the declared 1 nm vertical resolution of the confocal microscope. The uncertainty contribution introduced by the thermal deformation of the samples was neglected since the measurements were performed inside a metrologic laboratory with a 20 °C ± 0.5 °C controlled temperature.

In order to evaluate the uncertainty related to the measurements of the actual measurands, one more source of uncertainty was taken into account: $u_{Sa,\,mill}$, the standard deviation of ten repeated *Sa* measurements on the micromilled samples and on their replicas. Therefore, the expanded uncertainty of surface roughness measurement of the micromilled surfaces is calculated as:

$$U_{Sa} = k \times \sqrt{u^2_{CONF} + u^2_{Sa,\,mill}} \tag{2}$$

where *k* is the coverage factor, equal to 2 for a 95% confidence interval. In order to characterize the uncertainty of both masters and replicas, $u_{Sa,\,mill}$ was calculated for the three materials involved in the study. Table 4 presents the uncertainty budget.

Table 4. Uncertainty contributions for the *Sa* roughness measurements on the masters and replicated surfaces.

Uncertainty Contribution	AISI 440 H	AISI 440 A	Silicone
u_{cal}/nm	11.0	11.0	11.0
u_p/nm	2.1	2.1	2.1
$u_{res,CONF}$/nm	0.3	0.3	0.3
u_{CONF}/nm	11.2	11.2	11.2
$u_{Sa,\,mill}$/nm	3.7	3.9	4.7
U_{Sa}/nm	24	24	24

The role of U_{Sa} is fundamental in the replication performance verification: if the uncertainty intervals associated with direct and the indirect measurements overlap, the two outputs must be considered as equal. It is worth noting that the preponderant uncertainty contribution is represented by the calibration certificate of the roughness standard artifact.

2.5. Geometrical Measurement Methodology and Uncertainty Evaluation Procedure

The widths measurements of the micromilled pockets were carried out using a focus variation optical instrument (InfiniteFocus from Alicona Imaging GmbH, Raaba, Austria). This type of microscope is suitable for acquiring three-dimensional micro geometries but not for characterizing roughness of the order of tens of nanometers [37], which is typical of micromilled samples. Table 5 shows the main characteristics of this instrument.

Table 5. Focus variation instrument characteristics.

Objective Magnification	10×
Numerical aperture	0.30
Working distance in mm	16.0
Field of view in μm	1429 × 1088
Digital lateral resolution in μm	0.88
Vertical resolution in nm	200.0

To evaluate the two measurands *W1* and *W2* (see Figure 2), the area corresponding to the whole pocket was acquired. Successively, *W1* and *W2* were extrapolated utilizing cross-sectional profiles (see Figure 4). In particular, one thousand parallel profiles were extrapolated and averaged and then the width was calculated as horizontal distance between the two points corresponding to the upper edge of the pocket vertical walls.

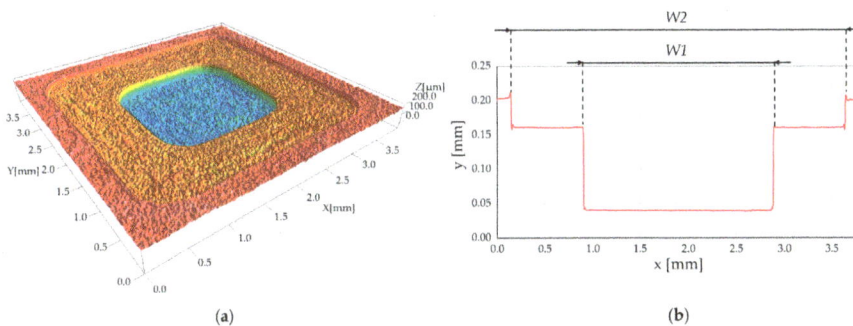

(a)   **(b)**

Figure 4. (**a**) Acquired 3D micromilled pocket (AISI 440 H, Pocket 3); and (**b**) extracted average profile and measurands *W1* and *W2*.

The uncertainty was evaluated using the same method applied for surface roughness measurements. In this case, a calibrated gauge block of 1.5 mm was selected as calibrated artifact. Therefore, the measurement uncertainty related to the width measurements using the focus variation instrument u_{FV} was calculated as:

$$u_{FV} = \sqrt{u^2_{cal} + u^2_p + u^2_{res,FV}} \tag{3}$$

where u_{cal} is the standard calibration uncertainty of the calibrated gauge and $u_{res,FV}$ is the resolution uncertainty related to the 2.0 μm lateral resolution of the focus variation microscope.

As in the previous case, the final expanded uncertainty was determined by adding the contributions of the micromilled samples. Thus, the total uncertainty associated with the measurement of the pocket width was calculated as:

$$U_W = k \times \sqrt{u^2{}_{FV} + u^2{}_{W,\,mill}} \tag{4}$$

where k is the coverage factor, equal to 2 for a 95% confidence interval and $u_{W,\,mill}$ is the standard deviation of ten repeated width measurements on the micromilled metal and replicated samples. As before, this last contribution was determined for the three materials under investigation. As the standard deviation on *W1* and *W2* measurements was equal for each material, U_W was applied to the measurements of both the measurands. Table 6 reports the resulting uncertainty budget.

Table 6. Uncertainty contributions for *W1* and *W2* measurements on the masters and replicated samples.

Uncertainty Contribution	AISI 440 H	AISI 440 A	Silicone
$u_{cal}/\mu m$	0.5	0.5	0.5
$u_p/\mu m$	0.7	0.7	0.7
$u_{res,FV}/\mu m$	0.6	0.6	0.6
$u_{FV}/\mu m$	1.0	1.0	1.0
$u_{W,\,mill}/\mu m$	1.2	1.1	1.5
$U_W/\mu m$	3.1	3.0	3.6

3. Results and Discussion

3.1. Comparison of Surface Topography Measurements

The replicated surfaces generally showed a good resemblance with the masters. The surface marks created by the micro tool were reproduced with fidelity, even in their finest details (see Figure 5). Moreover, the silicone accurately reproduced local discontinuities in the surface texture.

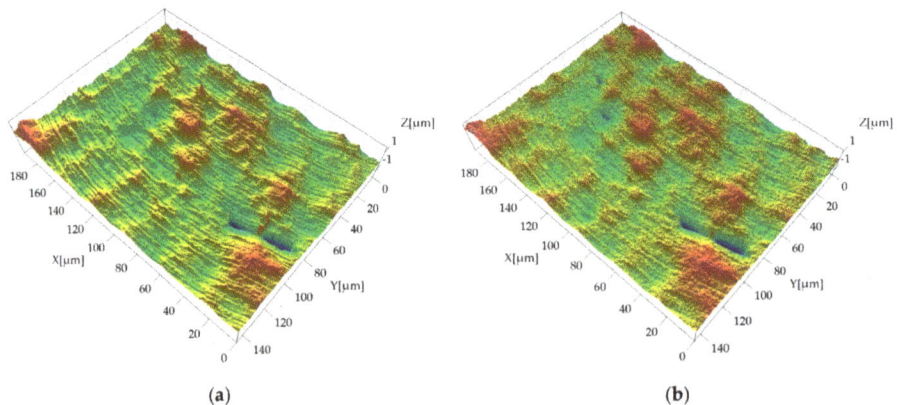

(a)   (b)

Figure 5. Detail of measured surface topography for AISI 440 H, surface S1, Sample 1: (**a**) metal master; and (**b**) silicone replica. The original silicone acquisition was inverted with respect to the *x*- and *z*-axes in order to facilitate the visual comparison. The *z*-axis was 20× magnified with respect to *x*- and *y*-axes.

The results of the quantitative surface characterization for the hardened mold steel are presented in Figure 6. For both Sample 1 and Sample 2, the direct measurement showed a decreasing average surface roughness when moving from the full slot surface (S1) to the other ones (S2 and S3). This is according to expectations, as S1 was machined in the cutting conditions that are typical of roughing operations. In general, *Sa* measured on the master surfaces ranged between 55 nm and 96 nm. A certain roughness variability was observed between the direct measurements of the two samples. In particular, Sample 1 presented an 11 nm to 18 nm larger *Sa* compared to Sample 2. This variability is due to the micromilling process itself, since the surfaces were machined in the same cutting conditions on the two samples. In comparison to the direct measurement, the three replicas always presented a larger *Sa*. However, considering the measurement uncertainty, direct and indirect measurements provided, in all the cases, the same *Sa* values. In fact, the uncertainty intervals overlap, making the measurement output equal from a metrological point of view. The standard deviations among the three silicone replicas range between 3 nm and 11 nm, demonstrating good repeatability of the replication procedure. Moreover, taking into account the measurement uncertainty, the replication procedure always provided consistent *Sa* results.

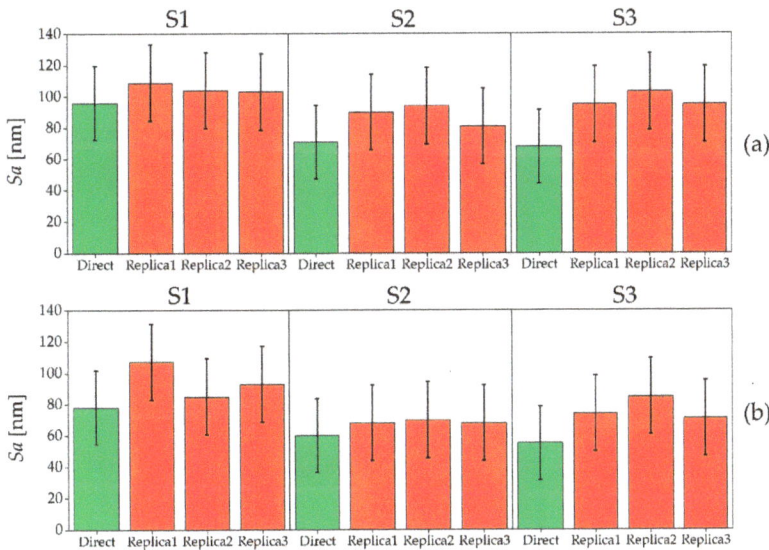

Figure 6. *Sa* values measured on master (green) and replicated (red) surfaces for AISI 440 H. The results for the three surfaces S1 (left), S2 (middle) and S3 (right) are shown: (**a**) Sample 1; and (**b**) Sample 2. The error bars indicate the expanded uncertainty U_{Sa}.

The results of the *Sa* measurements for the AISI 440 A are shown in Figure 7. In this case, S1 and S2 were produced with similar *Sa* roughness values (ranging between 61 nm and 74 nm for the two samples), while S3 had a significantly higher value of around 200 nm. This is contrary to expectations, as S3 was machined with the same radial depth of cut of S2. This unexpected increase of surface roughness was caused by the presence of plowing: for AISI 440 A, the last micromilling pass generated plowed marks on the surface, which diminished the surface quality (see Figure 8). This event happened only for S3 because the material accumulated on the tool during the first three passes (see Figure 1) leading to an increase of surface roughness [7] during the last pass. The softness of the annealed material is the most probable reason for this phenomenon. As for the previous material, a certain variability between Sample 1 and Sample 2 was observed. When comparing direct and indirect measurements, the replication process again introduced an average overestimation of the surface roughness of the master. However, as for the hardened material, the uncertainty intervals of direct and indirect measurements overlap in almost all cases, except for Replica 3 of S1, Sample 2 and Replica 2 of S3, Sample 1. The replication process is once again very repeatable: the three different replicas always provided the same measurements output, being the standard deviations among the three indirect measurements included between 7 nm and 15 nm.

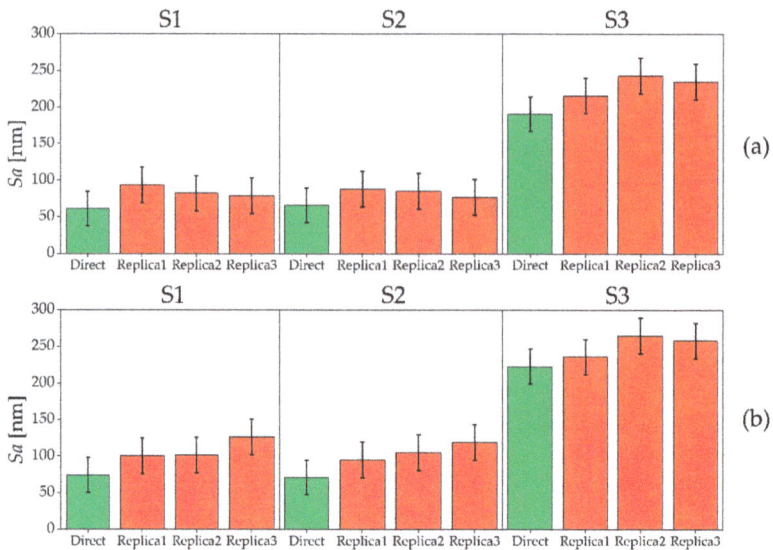

Figure 7. *Sa* values measured on master (green) and replicated (red) surfaces for AISI 440 A. The results for the three surfaces S1 (left), S2 (middle) and S3 (right) are shown: (**a**) Sample 1; and (**b**) Sample 2. The error bars indicate the expanded uncertainty U_{Sa}.

Figure 8. Detail of plowed marks on surface S3 of AISI 440 A material, Sample 2.

In both the hardened and annealed material, the replication procedure always introduced an overestimation of the roughness. The deviation Δ_{Sa}, calculated as the difference between average *Sa* of the three replicas and *Sa* of the master, did not show any dependence on the master roughness (see Figure 9). In fact, while the roughness of metal surfaces increased, the deviation Δ_{Sa} remained mostly constant around the value of 24 nm. The fact that the parameter Δ_{Sa} assumed similar values for all the measured surfaces also demonstrates that there was not a dependence of the replication performance with respect to the two samples, the two materials and the three surface types, as also shown by the ANOVA results in Table 7. The p-values were in fact always much larger than 5% for both single factors and two-ways interactions.

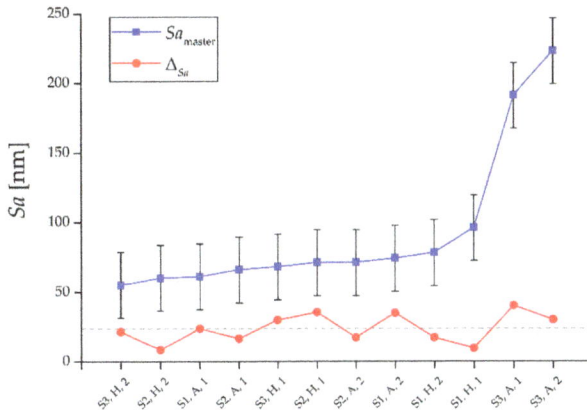

Figure 9. *Sa* values for the master surfaces and deviations $\Delta_{Sa} = Sa_{\text{replica}} - Sa_{\text{master}}$. The dashed grey line represents the average overestimation equal to 24 nm. The *x*-axis indicates the replicated surfaces by surface type (S1, S2 or S3), material (H or A) and Sample (1 or 2).

Table 7. ANOVA table for deviations Δ_{Sa} between replicated and master *Sa* surface roughness.

Factor	Adj. MS	F-Value	*p*-Value
Material	136.5	2.2	0.27
Surface type	137.7	2.3	0.31
Sample	57.7	1.0	0.43
Material × Surface type	115.8	1.9	0.34
Material × Sample	71.2	1.2	0.39
Surface type × Sample	145.2	2.4	0.30
Error	60.8		

To further investigate this experimental observation, other areal parameters were taken into account in order to understand which roughness component caused the *Sa* overestimation. In particular, the functional parameters *Svk*, *Spk* and *Sk* were calculated for masters and replicas and then compared. *Svk* is defined as the reduced dale height, and it provides an average indication of the valleys depth below the core roughness [33]. *Spk* is the reduced peak height and represents the mean height of peaks above the core surface [33]. Finally, *Sk* is used to characterize the core surface roughness [33]. In order to analyze how the silicone media replicated the valleys of the micromilled surfaces, it was chosen to compare the *Svk* parameter calculated on the masters with the *Spk* parameter of the corresponding replicas, since a valley on the master corresponds to a peak on the replica and vice versa. The core roughness was also utilized as parameter of comparison by means of *Sk* to determine whether the overestimation of the indirect *Sa* measurements is also related to an increase of this parameter.

The results of the comparison are shown in Figure 10. It is possible to observe that the *Svk* values of the master are systematically lower than the *Spk* values of the replicated surfaces. In particular, the *Spk* of the replicas is on average 60% higher than the *Svk* of the masters. This clearly demonstrates that the replication procedure generated surfaces having peaks that are higher than the valleys of the master. The same happens for *Sk*: the indirect measurements provided an average 63% higher level of *Sk* for all the produced surfaces, indicating that the core roughness is also increased after the replication procedure. Therefore, the *Sa* overestimation introduced by the indirect measurements is caused by two distinct phenomena: the increase of the height of the peaks with respect to depths of the valleys and the increase of the core roughness.

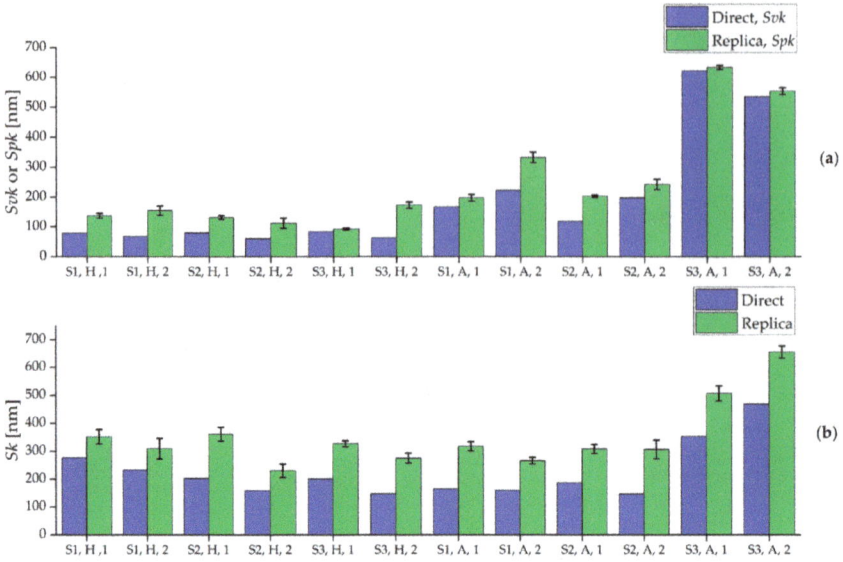

Figure 10. Functional roughness parameters calculated on master (blue) and replicated (green) samples: (a) *Svk* and *Spk* values for master and replicas respectively; and (b) *Sk* values. The interval bars indicate the standard deviations calculated among the three replicas. The *x*-axis indicates surface type (S1, S2 or S3), material (H or A) and Sample (1 or 2).

This phenomenon is probably generated during the demolding phase of the replica. The two-component silicone during solidification penetrates in the master surface valleys, replicating the surface topography (see Figure 5). When the solidified replica is manually removed, the silicone sticks to the deepest points of the surface valleys due to their very small width, making its removal more difficult than in other areas. Therefore, it appears that manual removal causes a nano-metric stretch perpendicular to the average plane of the surface, which results in the observed increase of both dale height and core roughness. This also explains why the overestimation is almost constant (see Figure 9): the manual removal acts as external factor, and it does not depend on the experimental variables. Another evidence is given by the Kurtosis parameter *Sku* [33], which provides a quantitative evaluation of the sharpness of the roughness profile. The replicated surfaces had, on average, a 23% higher *Sku* value, demonstrating that they have a sharper profile than the masters, as a consequence of the stretch induced by the manual removal.

3.2. Comparison of Geometrical Measurements

The steel micromilled pockets and their silicone replicas had a very similar shape. In particular, the replicas were able to accurately reproduce the geometry of the vertical walls, since no trace of any tilt due to the shrinkage of the polymer was present. With regard to the steps of the pockets, a not-perfect perpendicularity of the silicone pockets with respect to the vertical axis was sometimes observed, as shown in Figure 11 where Replica 3 presents a slightly tilted intermediate horizontal line. This micrometric deformation does not affect the width measurements.

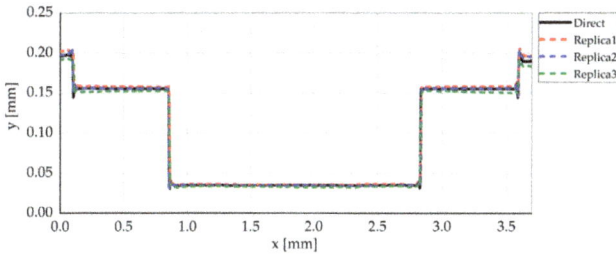

Figure 11. Extrapolated profiles for the direct and indirect width measurement. Original silicone acquisitions were inverted with respect to the *y*-axis in order to facilitate the visual comparison.

Figure 12 shows the results of the width measurements made on the AISI 440 H sample. The five steel micromilled pockets have an almost constant *W1*, while a certain variability was observed for *W2*. When comparing master and replicated widths, it is possible to observe that the indirect measurement generally provided a lower value for both the measurands. In fact, only in 10% of the cases the indirect evaluation generated an overestimation of *W1* and *W2*. This finding in is accordance with the experiments carried out by Madsen et al. [21], in which a shrinkage between master and PDMS replica was observed when measuring lateral geometrical features. Direct and indirect measurements provided the same output (i.e., the uncertainty intervals overlapped) in 53% of the comparisons for *W1* and in 40% for *W2*, revealing that the replication fidelity was higher for the measurement of the inner geometry of the pockets (see Figure 2). As regards the repeatability of the replication procedure, the average standard deviation, calculated among the three replicas and for the five pockets, equals 5 µm for *W1* measurements and 9 µm for *W2* measurements. It is therefore possible to conclude that the *W1* was replicated more accurately and precisely by the silicone media.

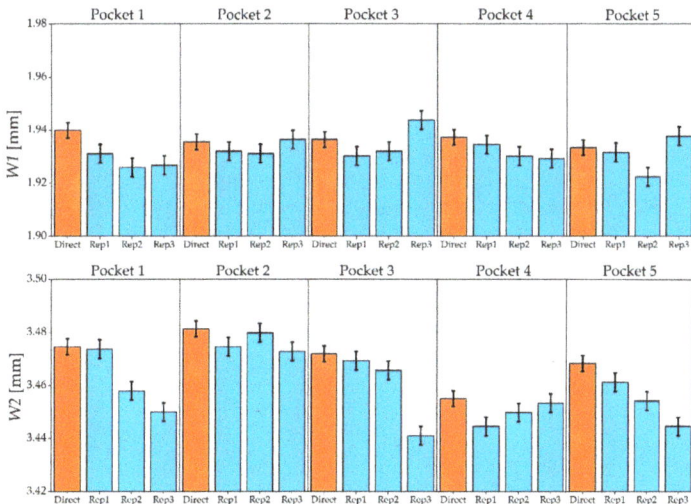

Figure 12. *W1* and *W2* measured on master (orange) and replicated (blue) pockets for AISI 440 H. The error bars indicate the expanded uncertainty U_W.

Figure 13 shows the results for the AISI 440 A sample. As in the harder material, the replication generally introduced an underestimation: only in one case out of 30 the width of the replicated specimen was larger than that of the master. The uncertainty intervals of direct and indirect measurements overlap in 67% of the cases for *W1* and in 47% of the cases for *W2*. Concerning the repeatability of the three replicas, average standard deviations of 3 μm and 8 μm were observed for *W1* and *W2*, respectively. Therefore, as for the hardened steel, a better replication was achieved for the internal width of the micromilled pockets.

The results in terms of deviation Δ_W, calculated as the difference between direct and indirect measurement outputs, are summarized in Figure 14 and Table 8 in which the ANOVA results for Δ_W are shown. It may be seen that average deviation was approximately 8 μm, demonstrating that the silicone media was able to achieve replication fidelity down to a single micrometer digit. The replication performance did not depend on the material of the master, as the p-value was larger than 5% for this experimental factor. The same conclusion can be drawn for the five pockets: the replication performance was unaffected by the different depth of the geometry under indirect measurement. On the contrary, the fidelity of the replicas was greatly affected by the type of measurand. When measuring *W2*, the deviation was on average 8 μm higher than when measuring *W1*. This suggests that the silicone media better replicates geometries that are more internal with respect to the outer surface on which it is applied (see Figure 2).

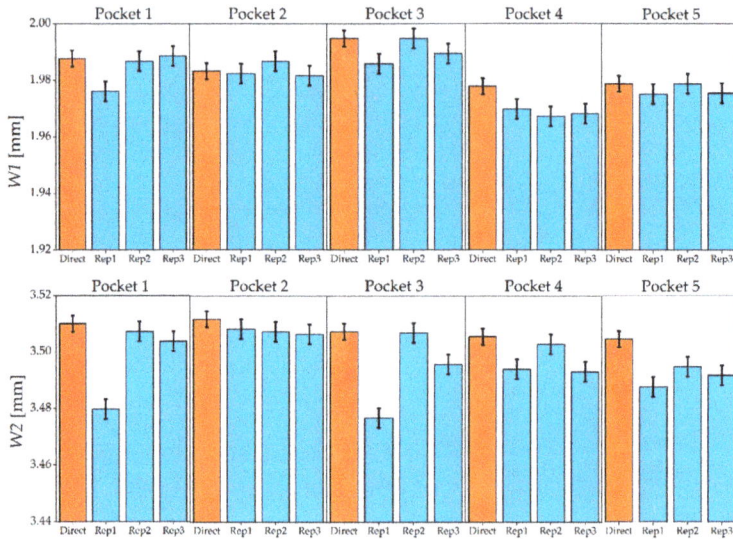

Figure 13. *W1* and *W2* measured on master (orange) and replicated (blue) pockets for the AISI 440 A. The error bars indicate the expanded uncertainty U_W.

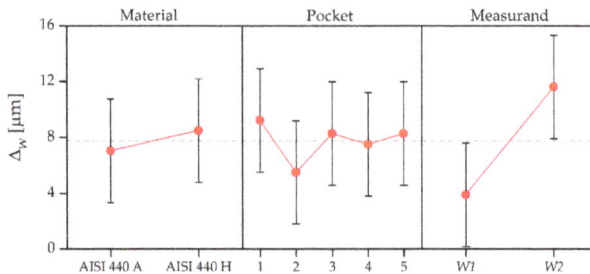

Figure 14. Main effects plot for the deviation $\Delta_W = W_{master} - W_{replica}$. The error bars indicate the measurement uncertainty of Δ_W, which was calculated applying the law of propagation of uncertainty [34].

Table 8. ANOVA table for deviations Δ_W between direct and indirect measurement of *W1* and *W2*.

Factor	Adj. MS	F-Value	*p*-Value
Material	32.2	0.6	0.46
Pocket	42.2	0.8	0.57
Measurand	800.2	14.1	0.00
Material × Pocket	56.9	1.0	0.42
Material × Measurand	5.6	0.1	0.76
Pocket × Measurand	69.8	1.2	0.31
Error	56.7		

In order to further characterize the behavior of the silicone when replicating geometrical features, the shrinkage s_W was calculated as:

$$s_W = \frac{W_{\text{master}} - W_{\text{replica}}}{W_{\text{master}}} \% \tag{5}$$

where W_{master} and W_{replica} are the width measurements carried out on the master and on the replica, respectively. By considering both *W1* and *W2*, the average shrinkage equaled 0.27% ± 0.03%, where the last value represents the expanded uncertainty calculated by means of the law of propagation of error [34] applied to Equation (5). This parameter is particularly useful, since it gives precise indications on what is the percentage underestimation introduced by the replication procedure when measuring micromilled geometries. Therefore, it can be used to determine the dimension of the master when only an indirect measurement is available, provided that the uncertainty value of the shrinkage is taken into account.

4. Conclusions

The present paper investigated the replication performance of a commercial silicone replica media applied as an indirect measurement tool for micromilled components. The replication capabilities related to surface and geometrical characterization were assessed separately by using two specifically designed micromilled samples. An uncertainty evaluation procedure based on ISO 15530-3 was used to compare the results provided by direct and indirect measurements.

The analysis of surface roughness, based on the three-dimensional areal parameter *Sa*, was carried out using a confocal microscope. Three different types of surface were machined on two mold steels and successively replicated. A qualitative comparison between masters and their replicas showed that the replication media was capable of accurately reproducing the appearance of the micromilled surface texture. A numerical comparison revealed that the indirect measurements always overestimated the average roughness of the master. By taking into account functional parameters such as *Svk*, *Spk*, and *Sk*, it was demonstrated that the overestimation introduced by the replication procedure is due to an increase of both core roughness and average peak height. The *Sa* deviation assumed an average value of 24 nm and was unaffected by the experimental variables (i.e., surface type and material). This strongly suggests that such effect was caused by an external factor such as the manual detachment of the silicone replicas from the master surface. However, considering the measurement uncertainty, the indirect and direct measurements provided the same results in 34 cases out of 36, demonstrating that the replication media used was suitable for characterizing micromilled surfaces that are inaccessible for other measurement systems.

The investigation related to the geometry replication focused on the measurement of widths of micromilled pockets featuring a two-stepped profile. Measurements were performed by means of a focus variation optical instrument. The tests were carried out on the same two materials employed in the previous analysis. Results showed that the replicated geometries were generally smaller than the metal masters. This underestimation was related to the type of measurand: for more internal geometries, such as the lower step of the pockets, the error was smaller, while it was larger for geometries that were more exposed to the pouring of the replication media. In view of the calculated expanded uncertainty, the direct and indirect width measurements provided the same result in 31 cases out of 60, demonstrating that the replication performance did not allow meeting the target consistently. However, considering the average shrinkage equal to 0.27% ± 0.03%, the master dimensions and the associated measurement uncertainty can be derived from indirect measurements performed with the investigated silicone media.

Future research will be dedicated to the study of different types of replication media and to micromilled components presenting more complex and free-form shapes.

Acknowledgments: This research work was undertaken in the context of MICROMAN project ("Process Fingerprint for Zero-defect Net-shape MICROMANufacturing", http://www.microman.mek.dtu.dk/). MICROMAN is a European Training Network supported by Horizon 2020, the EU Framework Programme for Research and Innovation (Project ID: 674801). Special thanks go to Eng. Marco Camagni and Pasquale Aquilino for their support in the experimental campaign and to Mahr company for providing the confocal microscope employed in the study.

Author Contributions: Federico Baruffi and Paolo Parenti conceived and designed experiments; Federico Baruffi, Paolo Parenti and Francesco Cacciatore performed experiments and measurements; Federico Baruffi and Guido Tosello analyzed the data; Federico Baruffi wrote the paper; and Massimiliano Annoni and Guido Tosello revised the paper.

Conflicts of Interest: The authors declare no conflict of interest.

References

1.   Alting, L.; Kimura, F.; Hansen, H.N.; Bissacco, G. Micro Engineering. *CIRP Ann.* **2003**, *52*, 635–657. [CrossRef]
2.   Brousseau, E.B.; Dimov, S.S.; Pham, D.T. Some recent advances in multi-material micro- and nano-manufacturing. *Int. J. Adv. Manuf. Technol.* **2010**, *47*, 161–180. [CrossRef]
3.   Bissacco, G.; Hansen, H.N.; De Chiffre, L. Micromilling of hardened tool steel for mould making applications. *J. Mater. Process. Technol.* **2005**, *167*, 201–207. [CrossRef]
4.   Hansen, H.N.; Hocken, R.J.; Tosello, G. Replication of micro and nano surface geometries. *CIRP Ann. Manuf. Technol.* **2011**, *60*, 695–714. [CrossRef]
5.   Parenti, P.; Masato, D.; Sorgato, M.; Lucchetta, G.; Annoni, M. Surface footprint in molds micromilling and effect on part demoldability in micro injection molding. *J. Manuf. Process.* **2017**, in press.
6.   Liu, X.; DeVor, R.E.; Kapoor, S.G.; Ehmann, K.F. The Mechanics of Machining at the Microscale: Assessment of the Current State of the Science. *J. Manuf. Sci. Eng.* **2004**, *126*, 666–678. [CrossRef]
7.   Weule, H.; Hüntrup, V.; Tritschler, H. Micro-Cutting of Steel to Meet New Requirements in Miniaturization. *CIRP Ann. Manuf. Technol.* **2001**, *50*, 61–64. [CrossRef]
8.   Hansen, H.N.; Carneiro, K.; Haitjema, H.; De Chiffre, L. Dimensional micro and nano metrology. *CIRP Ann. Manuf. Technol.* **2006**, *55*, 721–743. [CrossRef]
9.   Hocken, R.J.; Chakraborty, N.; Brown, C. Optical Metrology of Surfaces. *CIRP Ann. Manuf. Technol.* **2005**, *54*, 169–183. [CrossRef]
10. Leach, R.K. *Optical Measurement of Surface Topography*; Springer: Berlin, Germany, 2011.
11. Yao, D.; Kim, B. Injection molding high aspect ratio microfeatures. *J. Inject. Molding Technol.* **2002**, *6*, 11–17.
12. Mcfarland, A.W.; Poggi, M.A.; Bottomley, L.A.; Colton, J.S. Injection moulding of high aspect ratio micron-scale thickness polymeric microcantilevers. *Nanotechnology* **2004**, *15*, 1628–1632. [CrossRef]
13. Liou, A.C.; Chen, R.H. Injection molding of polymer micro- and sub-micron structures with high-aspect ratios. *Int. J. Adv. Manuf. Technol.* **2006**, *28*, 1097–1103. [CrossRef]
14. Bachmann, W.; Jean, B.; Bende, T.; Wohlrab, M.; Thiel, H.J. Silicone replica technique and automatic confocal topometry for determination of corneal surface roughness. *Ger. J. Ophthalmol.* **1993**, *2*, 400–403. [PubMed]
15. Scott, R.S.; Ungar, P.S.; Bergstrom, T.S.; Brown, C.A.; Childs, B.; Teaford, M.F.; Walker, A. Dental microwear texture analysis. *J. Hum. Evol.* **2006**, *51*, 339–349. [CrossRef] [PubMed]
16. Zuljan, D.; Grum, J. Non-destructive metallographic analysis of surfaces and microstructures by means of replicas. In Proceedings of the 8th International Conference of the Slovenian Society Non-Destructive Testing, Application of Contemporary Non-Destructive Testing in Engineering, Portoroz, Slovenia, 1–3 September 2005; pp. 359–368.

17. Jordon, J.B.; Bernard, J.D.; Newman, J.C. Quantifying microstructurally small fatigue crack growth in an aluminum alloy using a silicon-rubber replica method. *Int. J. Fatigue* **2012**, *36*, 206–210. [CrossRef]
18. AccuTrans 2012 AccuTrans Brochure. Available online: http://www.accutrans.info/fileadmin/dam/DATEN/AccuTrans/downloads/others/30000992_11--12_IFU_AccuTrans_AM_01.pdf (accessed on 15 February 2017).
19. Struers 2008 RepliSet Brochure. Available online: http://www.struers.com/en-GB/Products/Materialographic-analysis/Materialographic-analysis-equipment/Replication-system (accessed on 15 February 2017).
20. Tosello, G.; Hansen, H.N.; Gasparin, S.; Albajez, J.A.; Esmoris, J.I. Surface wear of TiN coated nickel tool during the injection moulding of polymer micro Fresnel lenses. *CIRP Ann. Manuf. Technol.* **2012**, *61*, 535–538. [CrossRef]
21. Madsen, M.H.; Feidenhans'l, N.A.; Hansen, P.-E.; Garnæs, J.; Dirscherl, K. Accounting for PDMS shrinkage when replicating structures. *J. Micromech. Microeng.* **2014**, *24*, 127002. [CrossRef]
22. Goodall, R.H.; Darras, L.P.; Purnell, M.A. Accuracy and precision of silicon based impression media for quantitative areal texture analysis. *Sci. Rep.* **2015**, *5*, 10800. [CrossRef] [PubMed]
23. Gasparin, S.; Hansen, H.N.; Tosello, G. Traceable surface characterization using replica moulding technology. In Proceedings of the 13th International Conference on Metrology Properties of English Surfaces, Kgs, Lyngby, Denmark, 12–15 April 2011; pp. 306–315.
24. Jinsheng, W.; Dajian, Z.; Yadong, G. A Micromilling Experimental Study on AISI 4340 Steel. *Key Eng. Mater.* **2009**, *408*, 335–338.
25. Kiswanto, G.; Zariatin, D.L.; Ko, T.J. The effect of spindle speed, feed-rate and machining time to the surface roughness and burr formation of Aluminum Alloy 1100 in micro-milling operation. *J. Manuf. Process.* **2014**, *16*, 435–450. [CrossRef]
26. Wang, J.; Gong, Y.; Shi, J.; Abba, G. Surface Roughness Prediction in Micromilling using Neural Networks and Taguchi's Design of Experiments. In Proceedings of the IEEE International Conference on Industrial Engineering and Engineering Managament (IEEM), Gippsland, Auatralia, 8–11 December 2009; pp. 1–6.
27. Cardoso, P.; Davim, J.P. Optimization of Surface Roughness in Micromilling. *Mater. Manuf. Process.* **2010**, *25*, 1115–1119. [CrossRef]
28. Liu, X.; DeVor, R.E.; Kapoor, S.G. Model-Based Analysis of the Surface Generation in Microendmilling—Part I: Model Development. *J. Manuf. Sci. Eng.* **2007**, *129*, 453–460. [CrossRef]
29. Abdelrahman Elkaseer, A.M.; Dimov, S.S.; Popov, K.B.; Negm, M.; Minev, R. Modeling the Material Microstructure Effects on the Surface Generation Process in Microendmilling of Dual-Phase Materials. *J. Manuf. Sci. Eng.* **2012**, *134*, 44501. [CrossRef]
30. Schwenke, H.; Neuschaefer-Rube, U.; Pfeifer, T.; Kunzmann, H. Optical Methods for Dimensional Metrology in Production Engineering. *CIRP Ann. Manuf. Technol.* **2002**, *51*, 685–699. [CrossRef]
31. Marinello, F.; Bariani, P.; De Chiffre, L.; Hansen, H.N. Development and analysis of a software tool for stitching three-dimensional surface topography data sets. *Meas. Sci. Technol.* **2007**, *18*, 1404–1412. [CrossRef]
32. MountainsMap®, Digital Surf. Available online: http://www.digitalsurf.fr/en/mntkey.html (accessed on 15 February 2017).
33. ISO. IOS 25178-2: *Geometrical Product Specifications (GPS)—Surface Texture: Areal—Part 2: Terms, Definitions and Surface Texture Parameters*; IOS: Geneva, Switzerland, 2012.
34. Joint Committee for Guides in Metrology (JCGM). *Evaluation of Measurement Data: Guide to the Expression of Uncertainty in Measurement*; JCGM: Paris, France, 2008.
35. ISO. IOS 15530-3: *Geometrical Product Specifications (GPS)—Coordinate Measuring Machines (CMM): Technique for Determining the Uncertainty of Measurement*; IOS: Geneva, Switzerland, 2011.
36. Haitjema, H. Uncertainty in measurement of surface topography. *Surf. Topogr. Metrol. Prop.* **2015**, *3*, 35004. [CrossRef]
37. Tosello, G.; Haitjema, H.; Leach, R.K.; Quagliotti, D.; Gasparin, S.; Hansen, H.N. An international comparison of surface texture parameters quantification on polymer artefacts using optical instruments. *CIRP Ann. Manuf. Technol.* **2016**, *65*, 529–532. [CrossRef]

micromachines

MDPI

Article

The Effects of Profile Errors of Microlens Surfaces on Laser Beam Homogenization

Axiu Cao [1,2], Hui Pang [1], Jiazhou Wang [1,2], Man Zhang [1], Jian Chen [2,3], Lifang Shi [1,*], Qiling Deng [1,*] and Song Hu [1]

[1]   Institute of Optics and Electronics, Chinese Academy of Sciences, Chengdu 610209, China; longazure@163.com (A.C.); wuli041@126.com (H.P.); ph@ioe.ac.cn (J.W.); zhangman881003@126.com (M.Z.); husong@ioe.ac.cn (S.H.)
[2]   University of Chinese Academy of Sciences, Beijing 100049, China; chenjian@mail.ie.ac.cn
[3]   State Key Laboratory of Transducer Technology, Institute of Electronics, Chinese Academy of Sciences, Beijing 100190, China
*   Correspondence: shilifang@ioe.ac.cn (L.S.); dengqiling@ioe.ac.cn (Q.D.); Tel.: +86-28-8510-1178 (L.S.); +86-28-8510-0032 (Q.D.)

Academic Editors: Hans Nørgaard Hansen and Guido Tosello
Received: 4 January 2017; Accepted: 8 February 2017; Published: 13 February 2017

Abstract: Microlens arrays (MLAs) are key optical components in laser beam homogenization. However, due to imperfect surface profiles resulting from microfabrication, the functionalities of MLAs in beam modulation could be compromised to some extent. In order to address this issue, the effects of surface profile mismatches between ideal and fabricated MLAs on beam homogenization were analyzed. Four types of surface profile errors of MLAs were modeled theoretically and numerical simulations were conducted to quantitatively estimate the effects of these profile errors on beam homogenization. In addition, experiments were conducted to validate the simulation results, revealing that profile errors leading to optical deviations located on the apex of microlenses affected beam homogenization less than deviations located further away from it. This study can provide references for the further applications of MLAs in beam homogenization.

Keywords: microlens array; beam homogenization; surface profile error; microfabrication

1. Introduction

Uniform illuminations on target surfaces are required in many applications such as laser fusion, laser cosmetology and material processing [1–3]. However, most laser beams are distributed in the Gaussian form or other nonuniform forms. Therefore, the shaping of an arbitrary input intensity into a top-hat format is a key issue [4,5].

Diffractive and refractive solutions have been used to improve the homogeneities of the laser beams. The diffractive elements are used for beam shaping, in which certain fractions of inputs are directed with specific angles, generating outputs with desired intensity distributions. Since diffractive solutions depend heavily on the light wavelength, they are only suitable for limited applications with low rates of energy utilization [6,7]. Meanwhile, the refractive elements are suitable for a wider spectral range due to lower dispersions compared to diffractive counterparts [8–10]. As a key component for light refraction, a microlens array (MLA) consisting of a Fourier lens and at least one regular microlens array has been used for beam homogenization. More specifically, the input radiation is firstly divided into multiple fractions by each channel of the tandem microlens array. Then, the fractions can be superposed with each other in the focal plane of the Fourier lens to form a light with a uniform distribution of intensities [9].

As a key optical component requesting high-accuracy surface profiles, the fabrication of MLAs is of importance. Thermal reflow is a well-established technique of fabricating MLAs by thermally flowing pre-patterned photoresist posts [11,12]. However, this approach cannot produce MLAs with high filling factors (~78% in an orthogonal array and ~90% in a hexagonal array). When MLAs fabricated by thermal reflow are used in beam homogenization, a zero-order spot with high intensities is always generated due to limitations in filling factors.

In order to address this issue, our group developed an approach based on moving-mask exposure, enabling the fabrication of MLAs with a filling factor of 100% [13]. However, due to the nonlinear effects in photoresist exposure, there is an issue of surface profile mismatches between the ideal and fabricated MLAs, leading to non-uniform intensity distributions in beam homogenization.

In this study, we evaluated the effects of surface profile errors on beam homogenization, including theoretical analysis, numerical simulations and experimental estimations. The manuscript is arranged as follows: Section 2 is the theoretical analysis where surface profile errors of MLAs are classified into four types. Section 3 is the numerical analysis where the beam deviations in response to four types of surface profile errors of MLAs are quantified. Section 4 is the experimental section where the results of numerical simulations are validated. Section 5 is the conclusion of this manuscript.

2. Principle

Most microlenses require a spherical surface where the sag height z_0 is given by Equation (1).

$$z_0 = \frac{cr^2}{1 + \sqrt{1 - c^2 r^2}} \tag{1}$$

where c is the curvature (the reciprocal of the radius), and r is the radial coordinate in lens units. However, fabrications always lead to errors of surface profiles, which can be classified into four types (see Figure 1). The symbol $\triangle$ represents the maximal profile mismatch between the designed and fabricated microlens. For type I and II errors, the maximal mismatches were located in the central part of the microlens (see Figure 1a,b), while for type III and IV errors (see Figure 1c,d), the maximal mismatches were located in about one-quarter or three-quarters of the microlens.

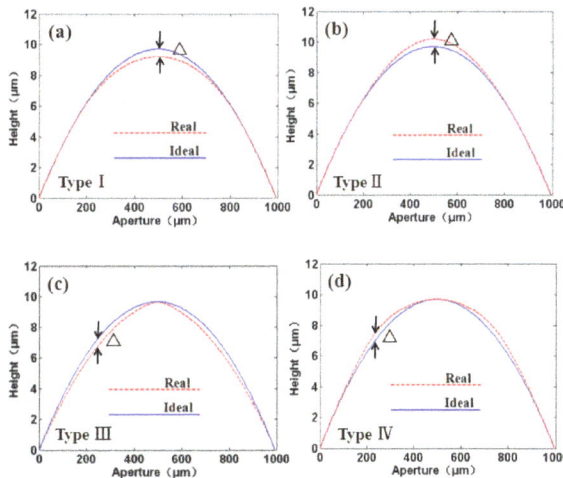

Figure 1. Four types of surface profile errors of microlenses: (**a**) type I; (**b**) type II; (**c**) type III; and (**d**) type IV. The dashed and solid lines represent the cross-section profiles of fabricated and ideal microlenses, respectively, where $\triangle$ represents the maximal profile mismatch between the ideal and fabricated microlenses.

Based on the numerical fitting of results from multiple experiments, surface profile errors z_i can be described as Equation (2).

$$z_i = \frac{cr^2}{1 + \sqrt{1 - c^2 r^2}} + \Delta_i, i = 1, 2, 3, 4 \tag{2}$$

where Δ_i ($i = 1, 2, 3, 4$) represents four types of surface profile errors which can be described in Equations (3)–(6), respectively.

$$i = 1, \Delta_1 = -\Delta \exp(\frac{-r^2}{w^2}) \tag{3}$$

$$i = 2, \Delta_2 = \Delta \exp(\frac{-r^2}{w^2}) \tag{4}$$

$$i = 3, \Delta_3 = -\Delta \sin(\frac{r}{d/2}\pi) \tag{5}$$

$$i = 4, \Delta_4 = \Delta \exp(\frac{-(r - \frac{d}{4})^2}{w^2}) \tag{6}$$

where w is the radius of the circle in which the maximal value of the surface profile error decreases to $1/e$ of the maximal value, and d is the aperture of the microlens.

3. Simulations

To analyze the effect of surface profile variations of MLAs on laser beam homogenization, numerical simulations were conducted. As shown in Figure 2, the simulations included a MLA and a Fourier lens. Based on the theory of scalar diffraction, when a laser beam transmits through a MLA and a Fourier lens in turn, the optical field at the focal plane of the Fourier lens is determined by the Fourier transformation of the transmission function of the MLA [14].

Firstly, the model for light field analysis and calculations based on the theory of scalar diffraction were proposed to study laser beam homogenization. The degree of beam homogenization based on the ideal spherical microlens was calculated by MATLAB (version 7.1, MathWorks, Natick, MA, USA). Then, spherical microlenses with surface profile errors were introduced to replace the ideal microlens to evaluate their effects on deviations of beam homogenization.

The key parameters used in the simulation were as follows. The light wavelength was 650 nm. The microlenses were tightly arranged in a hexagon for a filling factor of 100% (see Figure 2). The aperture of the microlens was hexagonal with a distance of 1 mm between the parallel edges. The focal lengths of the microlens and Fourier lens were 28 and 300 mm, respectively.

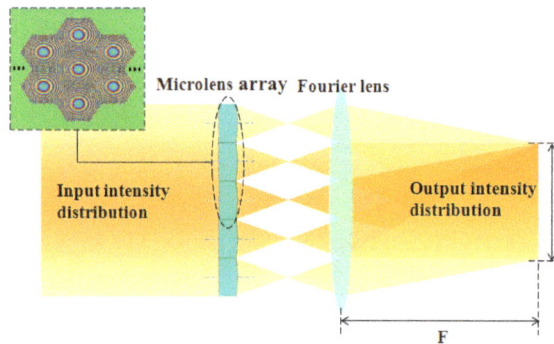

Figure 2. The numerical simulations of microlens array (MLA)-based beam homogenization, where an input optical beam transmits a MLA and a Fourier lens.

When the maximal surface profile errors were 0, 0.3, 0.5, 0.7 or 0.9 μm (the value of △ in microns has been quoted as a percentage of the microlens height), the corresponding deviation of beam homogenization was obtained, respectively (see Figure 3). With the increase of the surface profile errors, the side effects on the intensity distribution of the beam outputs became more and more obvious, indicated by a spot with non-uniform intensities. For type I and type IV errors, the output light was shown to gradually converge to the center along with the increasing values of the surface profile error. On the contrary, for type II and type III errors, the output light was shown to gradually deviate from the center along with the increasing values of the surface profile error. When the values of the surface profile errors were identical for these four types of situations, type III and IV errors exerted more influence than type I and II errors (see Figure 3c1–c4).

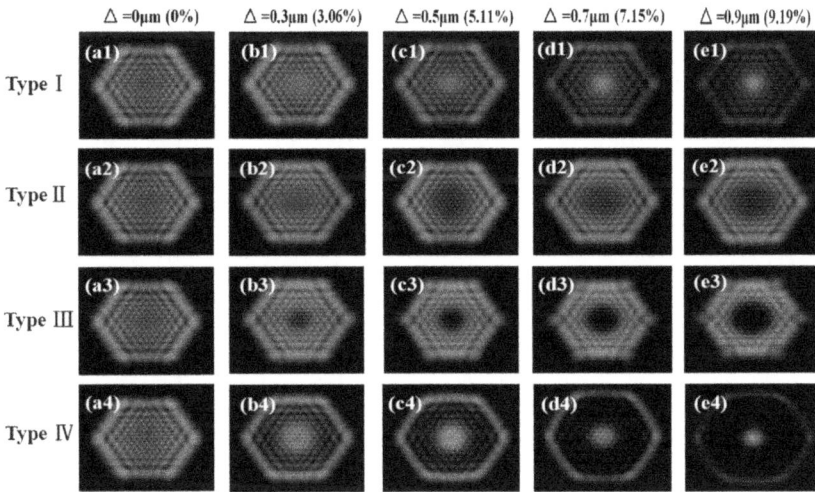

Figure 3. Numerical simulations of images after beam homogenization by passing MLAs with four types of surface profile errors. The maximal surface profile errors were 0 μm (0%) for (**a1**–**a4**); 0.3 μm (3.6%) for (**b1**–**b4**); 0.5 μm (5.11%) for (**c1**–**c4**); 0.7 μm (7.15%) for (**d1**–**d4**); and 0.9 μm (9.19%) for (**e1**–**e4**), respectively.

In order to quantitatively compare the influences of four types of surface profile errors on beam homogeneity, a key parameter η was defined as shown in Equation (7).

$$\eta = \frac{I_{\max} - I_{\min}}{I_{\max} + I_{\min}} \times 100\% \tag{7}$$

where $I_{\max}$ and $I_{\min}$ represent the maximal and minimal intensities within the output spot, neglecting the effect of the periodic interference pattern.

The non-uniformities of the surface profile errors within 1 μm were calculated in simulations, as shown in Figure 4. With an increase in the maximal surface profile error △, non-uniformity η was also shown to increase. Again, errors of type III and IV demonstrated more significant effects on beam homogeneity than the other two types of errors. More specifically, a sudden increase in the slope of the non-uniformity metric of 0.4 was observed for type III and IV errors and a further increase of errors led to rapid increases of non-uniformities.

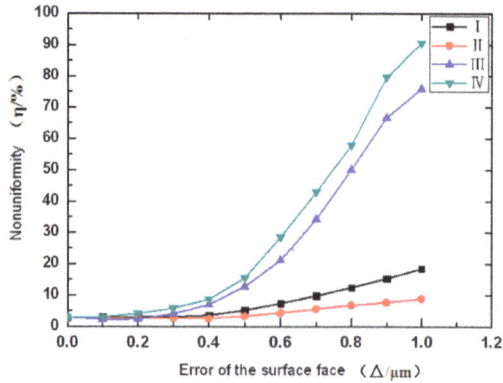

Figure 4. Non-uniformity η as a function of maximal surface profile errors (0–1 μm) of four types of profile mismatches.

4. Experiments and Discussion

In order to validate the simulation results, the corresponding experiments were conducted where silica was chosen as the substrate and AZ9260 (AZ Electronic Materials, Somerville, MA, USA) was used as the photoresist, which was spin-coated on the substrate at a speed of 3000 rpm for 20 s. Key parameters of the prebake temperature, the prebake period, and the obtained photoresist thickness were 100 °C, 30 min, and 9 μm, respectively. A gray-scale mask was fabricated to modulate the dose of exposure according to the surface profile of the MLAs (see Figure 5a). After exposure and development, photoresist-based MLAs were obtained. Then the etching of silica was conducted, which transferred the pattern of the MLAs to the silica substrate (see Figure 5b). A step profilometer (ALPHA-Step IQ, KLA-Tencor, Milpitas, CA, USA) was used to measure the surface profiles, validating the spherical structure of the MLAs (see Figure 5c).

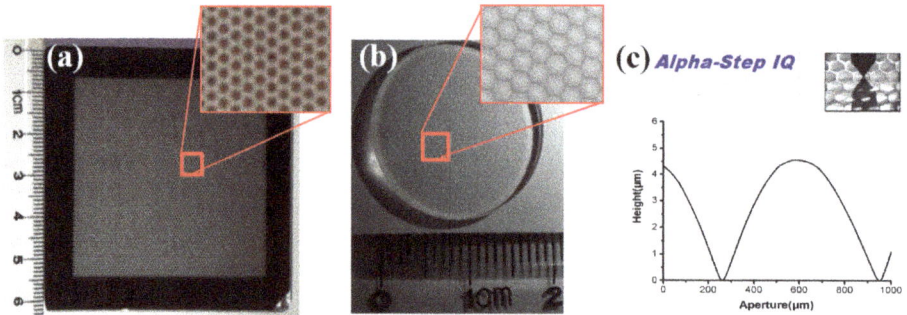

Figure 5. Key fabrication results, including (**a**) gray-scale mask; (**b**) prototype; (**c**) surface profiles.

A variety of MLAs fabricated with different surface profile errors were used to homogenize the beams, and the corresponding intensity distributions of beams after processing were captured and compared. The corresponding experimental results are shown in Figure 6, where the dashed and solid lines represented the cross-section profiles of fabricated and ideal microlenses, respectively. The maximal surface profile errors in Figure 6a–d were 0.55, 0.60, 1.78 and 0.32 μm, respectively. As to the error types, Figure 6a–d demonstrated four types of errors belonging to type I, II, III and IV, respectively. The theoretical boundary of the expected outer envelope of the irradiated region was drawn with lines in a white color on each image, as shown in Figure 6.

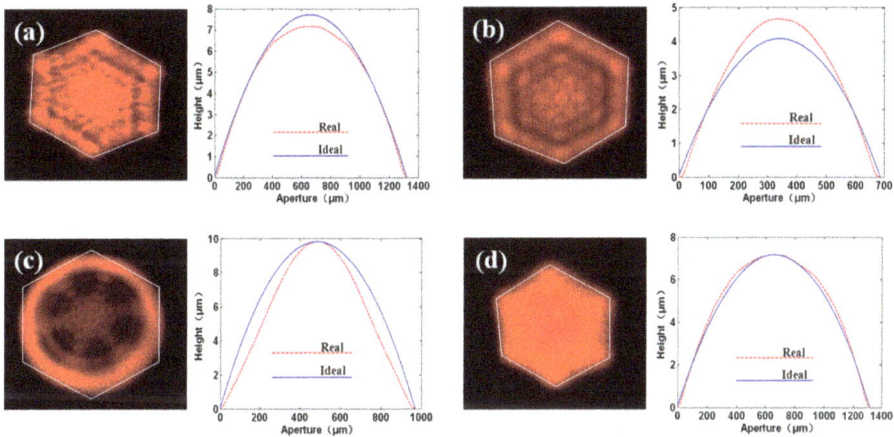

Figure 6. Experimental results of images processed by MLAs with different surface profile errors. The maximal errors were quantified as (**a**) $\triangle = 0.55$ μm (type I); (**b**) $\triangle = 0.60$ μm (type II); (**c**) $\triangle = 1.78$ μm (type III); and (**d**) $\triangle = 0.32$ μm (type IV).

This deviation of beam homogenization can be explained as follows. The input beam was firstly segmented into several sub-beams by the microlens and then the sub-beams were overlapped with each other through the Fourier lens, generating a uniform intensity distribution. The surface profile errors of the fabricated microlenses led to the irregular distributions of sub-beams, resulting in the redistribution of the divergence angles. Therefore, the output intensity was no longer uniform, and was affected by the fidelities of the surface profiles of microlenses.

If the fabricated microlens showed a type I, surface profile error the curvature of the surface profiles around the light axis at the central region of the microlens was shown to reduce. Therefore, the refractive angle of the light in the corresponding region decreased compared with the light in the same region of the ideal microlens. Since the scale of the light with smaller refractive angles increased, the intensity converged by the Fourier lens was significantly enhanced in the center of the output spot. On the contrary, if the fabricated microlens suffered from a type II surface profile error, the refractive angle of the light in the corresponding region was shown to increase because of the increase of the curvature of the surface profile. Therefore, the intensity was reduced in the center of the output spot because of the decreases in both light intensities and refractive angles.

In addition, when the fabricated microlens demonstrated a type III surface profile error, the curvature of the surface profile firstly increased and then decreased along the direction from the center to the edge of the microlens, compared with the ideal surface profiles. Then the amount of the light with smaller or bigger refractive angles decreased. Therefore, the intensity which should converge at the center and edge of the output spot was translated to regions between the center and the boundary areas. Thus, the intensities at the central and marginal regions were very weak. On the contrary, when the fabricated microlens showed a type IV surface profile error, the corresponding curvature of the surface profile changed in the opposite direction. Therefore, the intensity which should converge in the region between the central and the boundary areas of the output spot was modulated into the regions of the center and the edges.

Overall, the type III and IV surface profile errors nearly changed the whole surface of the microlens while type I and II surface profile errors only affected one half of the microlens. Therefore, the type III and IV surface profile errors demonstrated more significant influences on beam homogenization. In order to realize beam homogenization, the surface profile errors have to be controlled within a certain range at which the side influences can be tolerated. Meanwhile, the type III and IV surface profile errors should be avoided because of their significant side effects.

5. Conclusions

In conclusion, in this study, effects of surface profile mismatches between fabricated and ideal MLAs on beam homogenization were compared and studied. Both numerical simulations and experimental results quantitatively located the side effects of surface profile errors. These results suggest profile deviations located further away from the apex of the microlenses should be avoided in the future fabrication of MLAs since they demonstrated more significant side effects in beam homogenization in comparison to profile errors located on the apex of the microlenses. Hopefully, this study can provide references for the future study of beam homogenization–leveraging microlens arrays.

Acknowledgments: This research was supported by the National Natural Science Foundation of China (NSFC) (Nos. 61505214, 61605211); the Applied Basic Research Programs of Department of Science and Technology of Sichuan Province (Nos. 2016JY0175, 2016RZ0067); the National Defense Foundation of China (No. CXJJ-16M116); the Youth Innovation Promotion Association, Chinese Academy of Sciences (CAS); the CAS "Light of West China" Program. The authors thank their colleagues for their discussions and suggestions to this research.

Author Contributions: Axiu Cao and Hui Pang conceived and designed the experiments; Jiazhou Wang and Man Zhang performed the experiments while Lifang Shi and Qiling Deng supervised the whole process; Axiu Cao analyzed the data; Song Hu contributed reagents/materials/analysis tools; Axiu Cao and Jian Chen wrote and revised the paper.

Conflicts of Interest: The authors declare no conflict of interest.

References

1. Lou, S.; Zhu, H.; Han, P. Laser beam homogenizing system design for photoluminescence. *Appl. Opt.* **2014**, *53*, 4637–4644. [CrossRef]
2. Kim, J.W.; Mackenzie, J.I.; Hayes, J.R.; Clarkson, W.A. High-power Er: YAG laser with quasi-top-hat output beam. *Opt. Lett.* **2012**, *37*, 1463–1465. [CrossRef] [PubMed]
3. Yao, P.H.; Chen, C.H.; Chen, C.H. Low speckle laser illuminated projection system with a vibrating diffractive beam shaper. *Opt. Express* **2012**, *20*, 16552–16566. [CrossRef]
4. Sinhoff, V.; Hambuecker, S.; Kleine, K.; Ruebenach, O.; Wessling, C. Micro-lens arrays for laser beam homogenization and transformation. *Proc. SPIE* **2013**, *8605*, 860509.
5. Wippermann, F.; Zeitner, U.; Dannberg, P. Beam homogenizers based on chirped microlens arrays. *Opt. Express* **2007**, *15*, 6218–6231. [CrossRef] [PubMed]
6. Deller, M. *Laser Beam Shaping: Theory and Techniques*; Dickey, F.M., Holswade, S.C., Eds.; CRC Press (Taylor and Francis Group): Boca Raton, FL, USA, 2000.
7. Pepler, D.A.; Danson, C.N.; Ross, I.N.; Rivers, S.; Edwards, S. Binary-phase Fresnel zone plate arrays for high-power laser beam smoothing. *Proc. SPIE* **1995**, *2404*, 258–265.
8. Zhou, A.F. UV excimer laser beam homogenization for micromachining applications. *Opt. Photonics Lett.* **2011**, *4*, 75–81. [CrossRef]
9. Deng, X.M.; Liang, X.B.; Chen, Z.Z.; Yu, W.Y.; Ma, R.Y. Uniform illumination of large targets using a lens array. *Appl. Opt.* **1986**, *25*, 377–381. [CrossRef] [PubMed]
10. Kamon, K. Flye-Eye Lens Device and Lighting System Including Same. U.S. Patent No. 5,251,067, 5 October 1993.
11. Lin, C.P.; Yang, H.; Chao, C.K. Hexagonal microlens array modeling and fabrication using a thermal reflow process. *J. Micromech. Microeng.* **2003**, *13*, 775–781. [CrossRef]
12. Mohammed, A.; Cherry, G.; Franck, C.; Stuart, V.S.; Rajdeep, S.R. Geometrical characterization techniques for microlens made by thermal reflow of photoresist cylinder. *Opt. Lasers Eng.* **2008**, *46*, 711–720.

13. Shi, L.F.; Du, C.L.; Dong, X.C.; Deng, Q.L.; Luo, X.G. Effective formation method for an aspherical microlens array based on an aperiodic moving mask during exposure. *Appl. Opt.* **2007**, *46*, 8346–8350. [CrossRef] [PubMed]
14. Cao, A.X.; Pang, H.; Wang, J.Z.; Zhang, M.; Shi, L.F.; Deng, Q.L. Center off-axis tandem microlens arrays for beam homogenization. *IEEE Photonics J.* **2015**, *7*, 2400207. [CrossRef]

micromachines

MDPI

Article

Development of Novel Platform to Predict the Mechanical Damage of a Miniature Mobile Haptic Actuator

Byungjoo Choi [1], Jiwoon Kwon [2], Yongho Jeon [1] and Moon Gu Lee [1,*]

[1] Department of Mechanical Engineering, Ajou University, Suwon-si 16499, Korea; dasom@ajou.ac.kr (B.C.); princaps@ajou.ac.kr (Y.J.)

[2] Department of Convergence Technology Research, Korea Construction Equipment Technology Institute, Gunsan-si 54004, Korea; jwhj0814@gmail.com

* Correspondence: moongulee@ajou.ac.kr; Tel.: +82-31-219-2338

Academic Editors: Hans Nørgaard Hansen and Guido Tosello
Received: 28 February 2017; Accepted: 9 May 2017; Published: 13 May 2017

Abstract: Impact characterization of a linear resonant actuator (LRA) is studied experimentally by a newly-developed drop tester, which can control various experimental uncertainties, such as rotational moment, air resistance, secondary impact, and so on. The feasibility of this test apparatus was verified by a comparison with a free fall test. By utilizing a high-speed camera and measuring the vibrational displacement of the spring material, the impact behavior was captured and the damping ratio of the system was defined. Based on the above processes, a finite element model was established and the experimental and analytical results were successfully correlated. Finally, the damage of the system from impact loading can be expected by the developed model and, as a result, this research can improve the impact reliability of the LRA.

Keywords: drop test; impact analysis; reliability; haptic actuator; linear resonant actuator (LRA)

1. Introduction

Haptic perception can be classified into kinesthesia and tactility. The kinesthesia senses the mass, hardness, and shape of an object, whereas tactility senses the roughness, protuberance, and temperature of an object surface [1]. The haptic actuators which can communicate between a human and a machine based on the sense of touch are being developed. Several studies have recently focused on how to make this technology more realistic and immersive. Therefore, its field of application is expected to expand further.

Many studies have carried out the technology related to haptic perception over recent decades. The use of an eccentric rotating motor (ERM) as a vibrational source is reduced because of a lack of sensuous delivery caused by its slow response and narrow frequency range [2,3]. The solenoid resonant actuator (SRA) and piezoelectric resonant actuator (PRA) are introduced as having fast and wide frequency responses. However, the SRA is 25 mm or longer, making it too bulky to be applied to a small device, and the PRA requires a high-voltage amplifier and lacks structural durability because of the occurrence of the piezoelectric effect at high voltages and their embrittlement. In recent years, consequently, the LRA, with good durability and reasonable operating voltage, has been used by employing a mechanical spring. It can achieve the fast response and wide vibrational frequency range by using a small-sized voice coil motor (VCM) [4]. Generally, haptic actuators are constantly applied to mobile devices, medical instruments, automobiles, and entertainment devices. Additionally, high-definition (HD) haptic actuators are being developed for them. However, there are various impacts in the area where haptic actuators are applied. Due to these impacts, research is needed to make this device reliable. Such efforts have been made in related industries, but there is a need to make systematic forecasts because the existing ones are based on the trial and error method with the designer's experience.

Figure 1 shows the structure of an LRA (iPhone 4s, vibration motor). A spring, magnet, and moving mass are fixed on the top of the housing, while the coil is attached at the bottom of the housing. When a current is applied to the coil, a Lorentz force is generated due to the electromagnetic interaction between the current of the coil and the magnetic field. This force results in haptic perception by the vibration of the magnet and moving mass attached to the spring [5]. The structural durability of each component is a crucial part in the delivery of consistent haptic perception and, therefore, the related industry requires high reliability for this system. In the case of the impact durability test dropping from 1.8 m, from a human ear's height, only a 10% malfunction rate for the haptic actuator mounted in a smartphone is allowed in the industry. In order to satisfy this criterion, many suppliers have performed several research activities, such as reliability testing and analytical work [6–12]. The drop test, which is one of the experimental methods, is the most accurate way to study the impact resistance of a mobile device. However, experimental approaches cannot effectively characterize the impact behavior of each component because of its small size and behavior under a high-rate regime. In order to solve this issue, the finite element analysis was introduced in this study and this approach was verified by a comparison with the experimental approach.

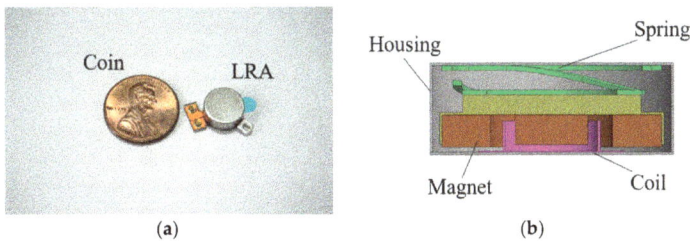

Figure 1. Appearance and internal structure of a coin-type LRA: (**a**) appearance; and (**b**) cross-sectional view.

Among the components in the actuator, the spring is the most sensitive part under impact loading and the impact damage can change the natural frequency of the spring, causing a malfunction, like insufficient acceleration.

In this study, we developed the novel platform which can predict the mechanical damage for LRA under impact loading. To generate the impact loading, the specific drop tester was developed and the repeatability test was executed for checking the feasibility of the test apparatus. Additionally, the analytical model was established and appropriate material testing was performed for obtaining the mechanical property simultaneously. As the relevance of the analytical model was proved by a correlation with experimental results, we can finally successfully predict the mechanical damage of the LRA under impact loading with the developed analytical model. We can assure that the research output explained in this article can play a significant role for damage analysis of various electrical devices under impact loading.

2. Drop Tester

2.1. Drop Tester Configuration

In general, the falling motion of the mobile device shows several aspects by rotation moment, air resistance, secondary impact, and so on. Therefore, it is very difficult to analyze the mechanical behavior experimentally due to the experimental uncertainty caused by the falling motion. In order to control the experimental uncertainty, a new drop tester was developed.

The experimental setup was constructed as shown in Figure 2. The impact force sensor (200C50, PCB Piezotronics Inc., Depew, NY, USA) can measure the frequency band of 0.0003 Hz to 30 kHz and a maximum dynamic compression force of 222.4 kN [13]. It was mounted onto the bottom of the test system. The measured signal was acquired and saved by a data acquisition system (LABVIEW, National Instruments, Austin, TX, USA). Additionally, the high-speed camera (FASCAM, APX-RS, Photron, Tokyo, Japan) was introduced for capturing the impact behavior visually and the image was captured at 512×512 resolution and 10,000 fps with an LED light.

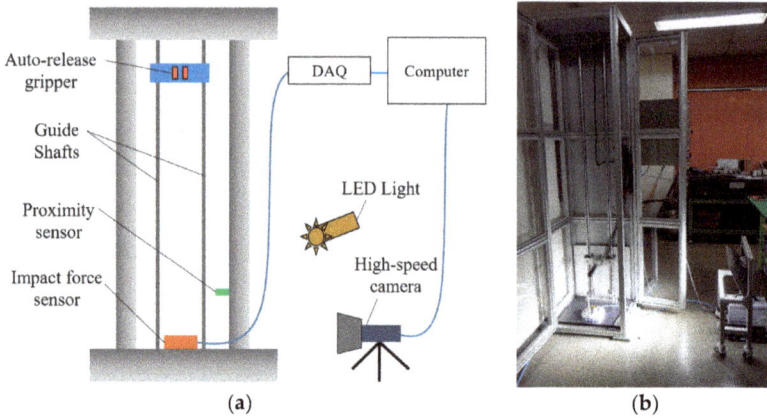

Figure 2. Drop tester configuration: (**a**) schematic diagram; and (**b**) experimental setup.

The test specimen was fixed with an auto-release gripper, which was operated by a pneumatic system, and the gripper was attached to the bushing. When the test begins, the bushing starts falling down through the guide shaft and then the gripper releases the specimen passing the proximity sensor. Finally, the test object impacts the force sensor mounted onto the bottom part of the impact tester. To reduce the experimental uncertainty, especially friction during the falling of the object, the bushing was composed of monomer cast nylon and grease was applied onto the guide shaft.

2.2. Drop Tester Verification

The developed drop tester consists of several supplementary components for controlling the experimental uncertainty and this supplementary component can distort the impact behavior. Therefore, the feasibility of this test apparatus must be verified in an appropriate manner. In this study, we fulfilled the verification of this feasibility by a comparison with the free fall test. A slender rod was used as a specimen and this rod was wrapped with silicone rubber to mitigate the vibration. Both the rod with and without the constraint were dropped from a height of 1.8 m. This height is a simulated value of the height of human ears. The drop from this height will result in a severe impact on the devices.

Despite the small allowable error in the peak force and period, as shown in Figure 3 and Table 1, the primary impact and subsequent oscillation for both tests was well matched and the agreement between the free and assisted tests suggests that the test results with the newly-developed test apparatus shows good repeatability.

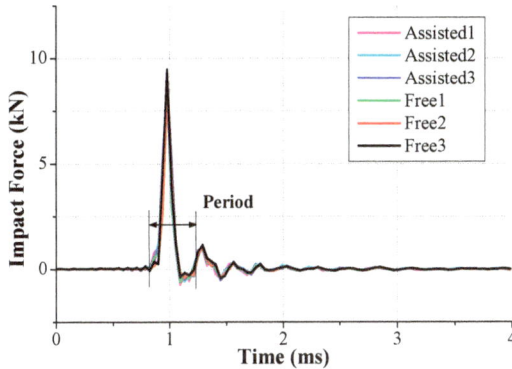

Figure 3. Comparison of the generated impact force between the free and assisted tests.

Table 1. The peak impact force and period from the free and assisted tests.

No.	Peak Impact Force (kN)	Period (ms)	Force Error (%)
Assisted1	8.68	0.39	2.36
Assisted2	8.84	0.47	0.56
Assisted3	9.23	0.47	3.82
Free1	8.78	0.39	1.24
Free2	8.31	0.43	6.52
Free3	9.48	0.43	6.64

For the comparison of the falling velocity just before the impact, the traveling time for a distance of 10 mm was measured with a high-speed camera, as shown in Figure 4. The traveling time was estimated as 1.7 ms for both tests. Therefore, we can conclude that the falling velocity is 5.88 m/s and this velocity can be used as the initial velocity for finite element analysis.

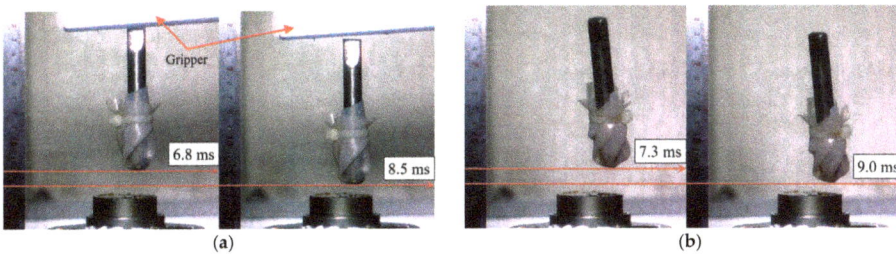

Figure 4. Comparison of falling velocity from the free and assisted tests: (**a**) assisted fall; and (**b**) free fall.

3. Finite Element Analysis of Impact Behavior of the LRA

3.1. Determination of Damping Ratio for the Spring

The damping ratio of the mechanical components is a major factor for determining the magnitude of the structural response under external loads. However, it is difficult to develop the constitutive model for structural damping because this is fully reliant on the dynamic condition. In this study, we used the experimental approach rather than an analytical approach to obtain the damping ratio of the spring. We applied a random excitation signal to the spring and extracted the vibrational peak (x_i)

and peak (x_{i+1}) at a certain time. The logarithmic decrement method is helpful to obtain the damping ratio by applying the extracted peaks to Equations (1) and (2) [14]:

$$\delta = \ln \frac{x_1}{x_2} \tag{1}$$

$$\zeta = \frac{\delta}{\sqrt{(2\pi)^2 - \delta^2}} \tag{2}$$

Figure 5 shows the experimental setup for measuring the vibrational behavior of the spring with a moving mass and the data was obtained by laser Doppler velocimetry (LDV). The direction of the vibration and excitation was matched with the falling direction of the dummy phone. Since the moving mass of the LRA is very small and the reflection characteristic is low, it is difficult to measure with laser. Then, the hexahedral magnet was attached to the upper part of the moving mass, so that the laser measurement focus was positioned on the side. The measurement was run on a vibration isolation stage (vibration isolation system, DAEIL SYSTEMS) to mitigate any vibrational noise and the data was acquired by an oscilloscope (Tektronix, Beaverton, OR, USA). As shown in Figure 6, the magnitude of the first peak was 0.255 and the second one was 0.225; therefore, the damping ratio (ζ) of the spring was eventually calculated as 0.02 using the logarithmic decrement method.

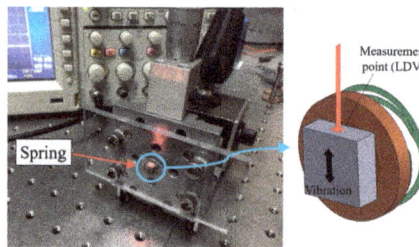

Figure 5. Experimental setup for measuring the vibrational behavior of the spring.

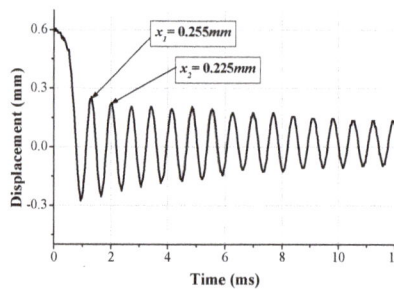

Figure 6. Measuring data for vibrational behavior.

3.2. Micro-Tensile Test for Spring

The LRA spring in this study is a thin plate 100 µm thick and 9 mm in diameter. The mechanical property of the thin plate is quite different from a bulk material because this can be changed by the correlation of the grain size with the component size [15]. Therefore, the micro-tensile test with a specially-prepared specimen was conducted to understand the mechanical property of the spring. The material of the specimen was SUS301 and this specimen was prepared with 200 µm thick, 3 mm gage length, and 19 mm total length using photo etching as shown in Figure 7. A universal material testing machine (UT-005, MTDI, Dajeon, Korea) was used in the micro-tensile test and the stress–strain curve of the SUS301 is shown in Figure 8. Even though the ultimate tensile strength (UTS) of SUS301 is known to be 1300 MPa [16], the UTS in this micro-tensile test was measured as 1510 MPa because of the size effect.

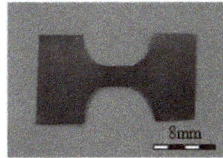

Figure 7. LRA spring test specimen.

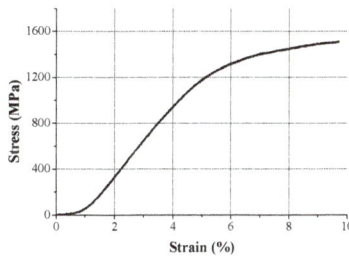

Figure 8. Stress–strain curve of SUS301 obtained by the micro-tensile test.

3.3. Experimental Verification of Analytical Model

To mimic the cellular phone, the appropriate dummy phone was prepared and the finite element (FE) model was also generated for this physical model, as shown in Figure 9. The dummy phone was composed of an aluminum panel and the LRA was attached to the center of the panel. In general, the LRA is covered with a metal housing. This makes it difficult to measure the vibration of the moving mass. Therefore, the LRA housing was removed to observe the moving mass movement during the impact moment. Additionally, this aluminum panel was covered by a transparent plate made of polycarbonate. This transparency allows the capturing of the behavior of the LRA by a high-speed camera. The hexa and tetra element was applied as the FE model element and a fine mesh with a 60 μm element size was applied in the spring and moving mass, which was our main interest. The impact that a haptic actuator dropped from the 1.8 m ear's height to the floor is a significant problem. In this case, a falling velocity was applied as 5.88 m/s, which was measured from the drop test just before collision, and gravitational acceleration was defined as 9.81 m/s². Futhermore, the friction coefficient was applied as 0.61 for static and 0.47 for kinetic, repectively.

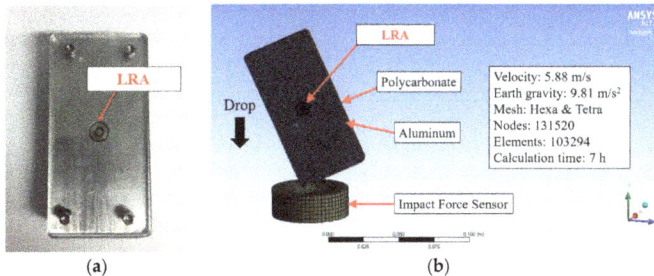

Figure 9. Physical and FE model of dummy phone with LRA: (**a**) dummy phone as a test specimen; and (**b**) finite element analysis.

We can identify the rebounding behavior of the moving mass from the high-speed camera and FE simulations, as shown in Figure 10. The movement was basically a relative motion of the dummy phone. The motion was measured by analyzing the high-speed image with the post-processing software Image J. A dot tracking method was applied to the center of the moving mass in the test. After numerical computation, tracking data was compared with the displacement of the center mesh from the finite element analysis (FEA) results. The moving mass was oscillated with 0.40 ms period in the drop test whereas the oscillating period was 0.24 ms in the FE simulation. Despite this discrepancy of the oscillating period, we can insist that the FE model was valid for this application because the moving mass in both cases was stabilized after two periods and the rate dependency of the metallic structure can be negligible.

To clearly understand the validity of the FE simulation, the traveling distance of the moving mass is presented in Figure 11. We can confirm that the trend of both cases was well matched even though there was no energy dissipation in the FE simulation caused by perfectly elastic modeling.

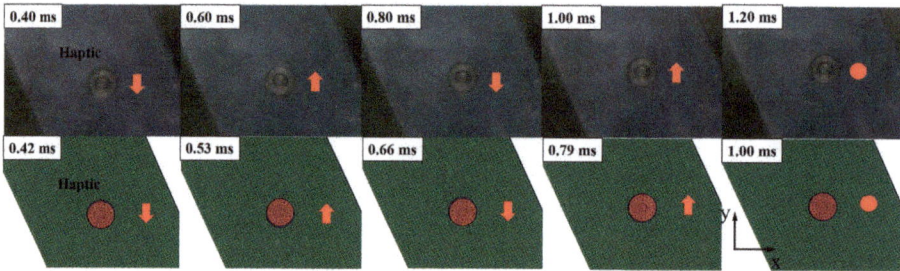

Figure 10. Impact behavior of the LRA in the drop test and FE simulations.

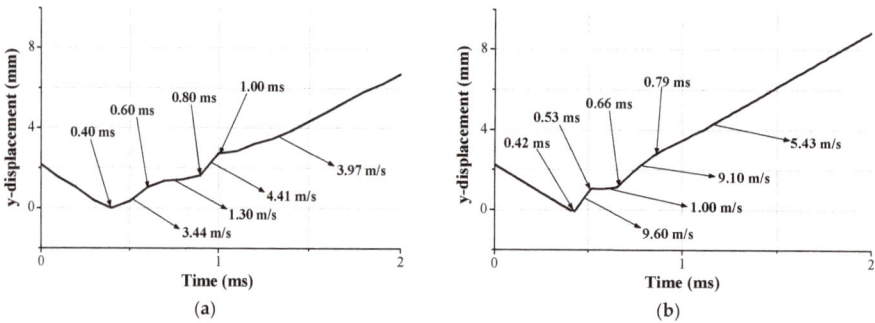

Figure 11. Comparison of traveling distance and moving velocity of the spring in the LRA: (**a**) drop test; and (**b**) FE simulation.

In spite of the short period of time, the estimation of the accuracy of the impact force is very important because it can cause external and internal damage to the structure. Therefore, the agreement of the impact force from the drop test and the FE simulation must be checked. The peak force from the drop test and the FE simulation was measured and verified the validity of the analytical model as shown in Figure 12 and Table 2. The four tests were conducted and the average peak force was measured as 3959.78 N with a 329.64 standard deviation. This value corresponds to the FE simulation results. On the other hand, it has the small error (7.5%) from drag and friction forces.

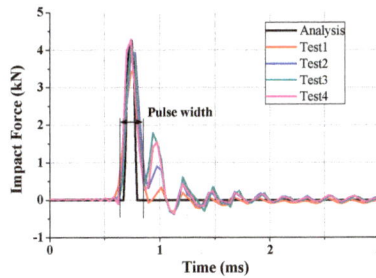

Figure 12. Impact force from the drop test and the FE simulation.

Table 2. Comparison of the impact force from the drop test and the FE simulation.

No.	Peak Force (N)	Pulse Width (ms)
Analysis	4255.94	0.083
Test1	3433.23	0.128
Test2	3934.92	0.126
Test3	4186.65	0.118
Test4	4284.30	0.121

3.4. FE Simulation

When a smart device, such as smartphone, drops on the floor, it starts the rotation due to rotational momentum generated by its asymmetric mass distribution. This rotation can increase the uncertainty and make appropriate analysis difficult. Therefore, the rotational motion of the dummy phone used in this study was restrained to avoid this uncertainty and to obtain the appropriate simulation results during the FE simulation. In the FE modeling, the vertical line on the ground was considered as the datum line, and the mass distribution of the dummy phone, which was the impacted object, was bilaterally symmetrized with respect to this datum line. At the rebounding state, we can achieve the translational motion of the impacted object without any rotational motion using this FE model. The stress contour from the FE simulation is shown in Figure 13. The impact occurred at 0.33 ms and maximum stress was observed as 4420 MPa at 0.39 ms. The stress was concentrated in the vicinity of the impact point and this concentration phenomenon was verified as observing the damage in the experiment. At 0.46 ms, the impacted object began rebounding and the LRA, including the moving mass and spring components, started the oscillation. The traveling behavior of the LRA spring attached to the impacted object is already shown in Figure 11b.

Figure 13. Stress (von Mises) contour from the FE simulation under impact loading.

In general, any component attached to a traveling object subjected to perpendicular motion is expected to travel along the same direction with the main object. In other words, the motion of the LRA can be expected to be in perpendicular motion to the ground in this study. In the FE simulation, however, the moving mass had asymmetrically oscillated to the datum line because of the asymmetricity of the spiral spring fixed to the moving mass. Therefore, the impact between the moving mass and the LRA housing was also asymmetric. Even though the first impact occurred perpendicular to the ground between element 3316 and the housing, all of the impact after the first one appeared as a rotational motion in the clockwise direction as shown in Figure 14. The impact stress of each element is shown in Figure 15 and the maximum stress was 699 MPa (0.43 ms), 515 MPa (0.51 ms), 293 MPa (0.65 ms), and 158 MPa (0.87 ms), respectively. This stress may cause spring damage.

In order to understand the spring behavior, the mechanical behavior of the spring was investigated. The stress and deformation state at a certain time is described in Figure 16. As mentioned above, the structural asymmetricity makes the vibration skew to the left. Due to the spring geometry, the stress was concentrated to the leg, which is in the vicinity of the supporting point, and the maximum stress was observed at elements 85,472, 86,187, and 91,428, corresponding to each leg. The downward maximum deformation of the spring occurred at 0.43 ms. At that instant, elements 85,472 and 86,187 were subjected to the tensile stress and the maximum stress was calculated as 623 MPa at element 85,472, whereas element 91,428 was subjected to the compressional stress and the stress value was observed as 203 MPa. On the other hand, the upward maximum deformation of the spring occurred at 0.51 ms. At that instant, the stress state of each element was reversed and the stress value was magnified by stress accumulation. The stress values of elements 85,472 and 86,187 were 1307 MPa and 974 MPa, respectively, and the stress value of element 91,428 was 1359 MPa. During the whole transient state, the maximum stress occurred in the upward bouncing of the second period. The measured value was 1695 MPa, which is over the UTS of the spring and, therefore, we can expect severe damage to the spring [17].

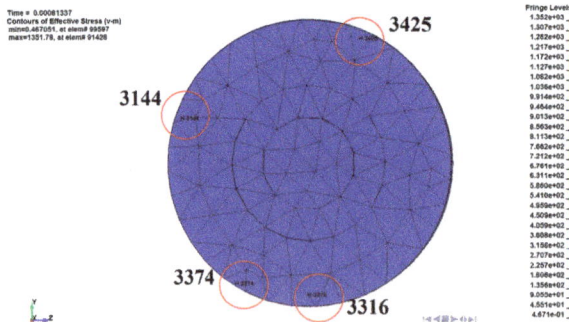

Figure 14. Contact element of the moving mass fixed to the spring in the LRA.

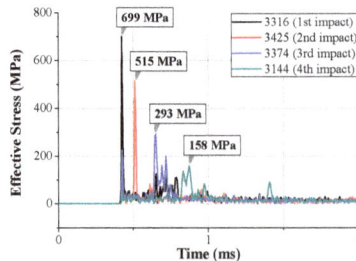

Figure 15. Impact stress (von Mises) generated in contact elements of the moving mass (from Figure 14).

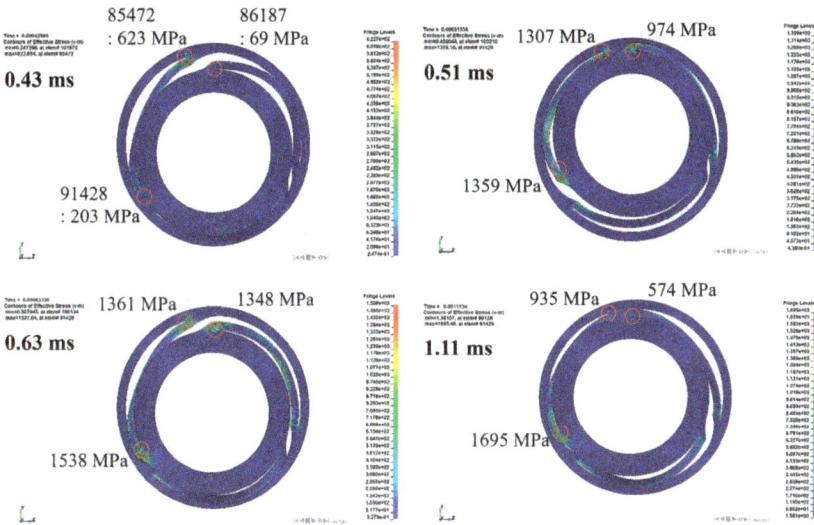

Figure 16. Impact stress (von Mises) and deformation of the spring in the LRA during impact loading.

Figure 17 illustrates the effective plastic strain for certain elements subjected to concentrated stress. The plastic deformation began at 0.35 ms as the impact was transmitted to the spring. The initial strain slope of each element was similar, but the strain was increased due to the high stress condition after the first period (0.63 ms). Even the strain of element 86,187 overtook the strain of element 85,472 at 1.0 ms because the accumulation and dissipation of the impact energy was different for each element during the transient state. The maximum strain was calculated as 0.135, 0.182, and 0.152 at element 85,472, 91,428, and 86,187, respectively.

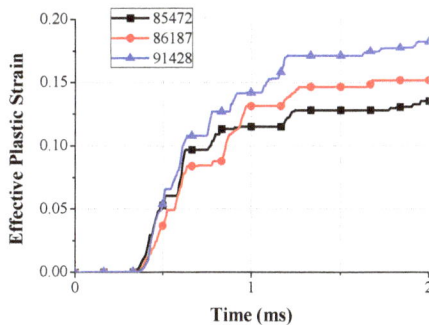

Figure 17. Effective plastic strain for certain elements in the spring during impact loading.

The expected deformation shape of the spring was calculated by FE simulation and this is illustrated in Figure 18. The upper plate, which is the spring leg, is severely deformed and, therefore, we can easily expect that the vibrational characteristic of the LRA is changed. This means that this LRA is no longer providing the haptic perception.

Figure 18. Expected deformation shape (highlighted with red lines) of the spring after the impact.

4. Conclusions

We developed a novel platform which can predict the mechanical damage for an LRA under impact loading by using drop test and FE analysis methods.

(1) For the analysis, a series of preparations were carried out. First, the drop tester was newly-developed for experimental verification of the FE model. Its experimental verification satisfied the free fall conditions while assisting the drop with a test apparatus. Second, a micro-tensile test was performed to obtain the material properties considering the size effect of the thin LRA springs. Third, structural damping was modeled by measuring the vibration displacement of a spring with the excitation signal.

(2) Based on the previous study, the impact FE modeling of a dummy phone including an LRA was performed, and its experimental verification was carried out by comparison of the impact deformation and force during the impact behavior. Despite the error in the impact force (7.5%) and pulse width (33%), the analytical model and experimental model were well correlated. Additionally, the impact rebound displacement is well matched.

(3) Consequently, the damage of the FE model was analyzed. The external impact and secondary internal impact of the LRA moving mass were concentrated on the LRA spring. Primary and secondary impact generated a maximum impact stress of 1695 MPa. Further, effective strain at the same position was evaluated as 0.182. The damaged shape of the spring was confirmed and a vibration characteristic change was expected.

In conclusion, impact deformation and force were calculated through an experimentally-verified FE model. This process can redeem the durability study against impact which has been conducted by the designer's experience and trial-and-error. Finally, this research will be used extensively in impact analysis of smart devices, automobiles, medical instruments, game machines, and remote controls, including miniature parts.

Acknowledgments: This work was supported by the National Research Foundation of Korea (NRF) grant funded by the Korea government (MISP) (no. 2014R1A2A10052344).

Author Contributions: M.G.L., Y.J., J.K. and B.C. conceived and designed the experiments and simulations; B.C. designed the drop tester and performed the experiments and simulations; Y.J., M.G.L. and B.C. analyzed the data; B.C., M.G.L. and J. K. wrote the paper.

Conflicts of Interest: The authors declare no conflict of interest.

References

1. Kern, T.A. *Engineering Haptic Devices*; Springer Heidelberg: Berlin, Germany, 2009.
2. Cho, Y.J.; Yang, T.H.; Kwon, D.S. A new miniature smart actuator based on piezoelectric material and solenoid for mobile devices. In Proceedings of the 5th International Conference on the Advanced Mechatronics (ICAM), Osaka, Japan, 4–6 October 2010; pp. 615–620.
3. Yang, T.H.; Pyo, D.B.; Kim, S.Y.; Cho, Y.J.; Bae, Y.D. A New Subminiature Impact Actuator for Mobile Devices. In Proceedings of the IEEE World Haptics Conference, New York, NY, USA, 21–24 June 2011; pp. 95–100.
4. Pyo, D.B. A Novel Impact-Resonant Actuator for Mobile Devices. Master's Thesis, Korea Advanced Institute of Science and Technology (KAIST), DaeJeon, Korea, 2012.

5. Kim, K.H.; Choi, Y.M.; Gweon, D.G.; Lee, M.G. A novel laser micro/nano-machining system for FPD process. *J. Mater. Process. Technol.* **2008**, *201*, 497–501. [CrossRef]
6. Goyal, S.; Buratynski, E. Methods for Realistic Drop-Testing. *Int. Microelectron. Packag. Soc.* **2000**, *23*, 45–52.
7. Kim, J.G.; Lee, J.Y.; Lee, S.Y. Drop/Impact Simulation and Experimental Verification of Mobile Phone. *Trans. Korean Soc. Mech. Eng.* **2001**, *25*, 695–702.
8. Kim, J.G.; Park, Y.K. Experimental Verification of Drop/Impact Simulation for a Cellular Phone. *Exp. Mech.* **2004**, *44*, 375–380. [CrossRef]
9. Zhou, C.Y.; Yu, T.X.; Lee, S.W. Drop/impact tests and analysis of typical portable electronics devices. *Int. J. Mech. Sci.* **2008**, *50*, 905–917. [CrossRef]
10. Karppine, J.; Li, J.; Pakarinen, J.; Mattila, T.T.; Paulasto-Krockel, M. Shock impact reliablity characterization of a handheld product in accelerated tests and use environment. *Microelectron. Reliab.* **2012**, *52*, 190–198. [CrossRef]
11. Mattila, T.T.; Vajavaara, L.; Hokka, J.; Hussa, E.; Makela, M.; Halkola, V. Evaluation of the drop response of handheld electronic products. *Microelectron. Reliab.* **2014**, *54*, 601–609. [CrossRef]
12. Choi, B.J.; Yeom, H.H.; Jeon, Y.H.; Lee, M.G. Experimental verification of drop impact test and analysis for mobile electronics. In Proceedings of the 11th International Conference on Multi-Material Micro Manufacture (4M) and the 10th International Workshop on Microfactories (IWMF), Kongens Lyngby, Denmark, 13–15 September 2016; pp. 125–128.
13. Metz, R. Impact and Drop Testing with ICP Force Sensors. *Sound Vib.* **2007**, *11*, 18–20.
14. Inman, D.J. *Engineering Vibration*; Pearson Education: London, UK, 2008.
15. Bazant, Z.P. Size effect on structural strength: A review. *Arch. Appl. Mech.* **1999**, *69*, 703–725.
16. Gupta, R.K.; Mathew, C.; Ramkumar, P. Strain Hardening in Aerospace Alloys. *Front. Aerosp. Eng.* **2015**, *4*, 1–13. [CrossRef]
17. Magd, E.E. Mechanical properties at high strain rates. *J. Phys. IV Colloq.* **1994**, *4*, 149–170.

micromachines

MDPI

Article

Modeling of the Effect of Process Variations on a Micromachined Doubly-Clamped Beam

Lili Gao, Zai-Fa Zhou * and Qing-An Huang *

Key Laboratory of MEMS of the Ministry of Education, Southeast University, Nanjing 210096, China;
LilyGaoChina@gmail.com
* Correspondence: zfzhou@seu.edu.cn (Z.-F.Z.); hqa@seu.edu.cn (Q.-A.H.);
  Tel.: +86-25-8379-2632 (Z.-F.Z.); +86-25-8379-8801 (Q.-A.H.)

Academic Editors: Hans Nørgaard Hansen and Guido Toselloame
Received: 1 October 2016; Accepted: 28 February 2017; Published: 5 March 2017

Abstract: In the fabrication of micro-electro-mechanical systems (MEMS) devices, manufacturing process variations are usually involved. For these devices sensitive to process variations such as doubly-clamped beams, mismatches between designs and final products will exist. As a result, it underlies yield problems and will be determined by design parameter ranges and distribution functions. Topographical changes constitute process variations, such as inclination, over-etching, and undulating sidewalls in the Bosch process. In this paper, analytical models are first developed for MEMS doubly-clamped beams, concerning the mentioned geometrical variations. Then, finite-element (FE) analysis is performed to provide a guidance for model verifications. It is found that results predicted by the models agree with those of FE analysis. Assigning process variations, predictions for performance as well as yield can be made directly from the analytical models, by means of probabilistic analysis. In this paper, the footing effect is found to have a more profound effect on the resonant frequency of doubly-clamped beams during the Bosch process. As the confining process has a variation of 10.0%, the yield will have a reduction of 77.3% consequently. Under these circumstances, the prediction approaches can be utilized to guide the further MEMS device designs.

Keywords: doubly-clamped beam; process variations; FE analysis; Bosch process; yield prediction

1. Introduction

Precise processing control has turned into an issue, owing to the mass production of micro-electro-mechanical systems (MEMS) devices and their increasingly complicated manufacturing processes. Discrepancies between initial designs and products deteriorate quickly with the feature size reductions. Even with the state-of-art fabrication techniques, process variations occur inevitably [1–3]. The process variations mainly include misalignment, footing as well as critical dimension (CD) loss [4], manifested as inclination, over-etching and undulating sidewalls in the Bosch process [5]. Typically, the effects of relative tolerances in MEMS devices are more severe than macro-scale products [1,6,7]. Relative manufacturing tolerances are alternatives to performance uncertainties. In addition, microstructures are commonly performed with nonlinear parallel plate electrostatic forces, which contributes to the complexity of problems. For such a reason, adequate methods are required due to the limitations in design rules and linear theories of MEMS [8].

Studies on process variations have achieved abundant results [9] in the scope of integrated circuits (ICs). However, these IC achievements cannot meet the whole needs of MEMS technologies. The conventionally used trial-and-error approach severely relies on the design-test cycle, which not only postpones the development cycle, but is also costly and time-consuming. Therefore, efforts have been made in the domain of MEMS devices analysis, for a better understanding of the impacts of process variations and also for reductions of performance variabilities at the design stage [10–15].

Islam et al. [10] conducted simulations and stress analysis on a fixed-fixed beam in electrostatic situations. Their results have reflected that changes in length and thickness tend to be more strictly controlled. Microbeam resonators are commonly utilized to detect or filter signals in MEMS. Due to manufacturing uncertainties, microbeam resonators undergo significant variability from initial designs. For example, Liu et al. [11] achieved tradeoff designs regarding multiple and conflicting design criteria, while Rong et al. [12] focused on multilayer structures while considering the first and second-order sensitivities of frequency. Mawardi et al. [13] utilized enumeration search and input–output relationships to get the governing parameters as well as a wide range for operating resonant frequency. In addition, magnetometers adopted multiphysics-based optimization and nonlinear situations [14], and gyroscopes focused on the packaging with double yield [15]. Except for unique device analysis, methods that are generally applicable have advanced the processing improvement further [3,16–26]. Mirzazaden et al. [16–18] investigated morphology uncertainties with reduced-order models through on-chip tests. Moreover, Shavezipur et al. [3,19], and Allen et al. [20] proposed the first-order second-moment (FOSM) and advanced FOSM reliability method, respectively, in a probabilistic way to obtain a linearized feasible region and maximize the yield. For those non-linear actuated MEMS devices, high fidelity optimization schemes have also been realized. To avoid the brute-force Monte Carlo (MC) scheme, Pfingsten et al. [21] considered a Bayesian Monte Carlo approach for yield estimation, with 90.0% computational savings but the same accuracy, compared with MC schemes. Vudathu et al. [22,26] applied a sensitivity analyzer for MEMS (SAM) by worst-case analysis, revealing the effects of parametric variations on performance and yield. Achievements have been reached to get a balance between precision and calculation—for example, the Sigma-Point approach applied to MEMS resonators with four orders of magnitude faster than MC [23], the generalized polynomial chaos (GPC) framework to handle stochastic coupled electromechanical analysis with the same precision and one order of magnitude faster compared with MC [24], and the Taguchi parameter design and statistical process-control method to minimize variability in performance response to fluctuations [25].

However, little work has been carried out for detailed analysis of specific processing steps. This paper intends to explore the effect of process variations on the resonance frequency of doubly-clamped beams, under the Bosch target processing environment. The commercial code ANSYS (11.0) [27] guides the verifications of the presented methods. The results suggest that with assigned process variations, structure performance and yield can be predicted. On the other hand, given design specifications, reasonable suggestions can be made for parameter error ranges under process variations.

2. Process Variations

Marked as a highly anisotropic etching process with high aspect ratios, deep reactive-ion etching (DRIE) is employed to create deep penetration, steep-sided holes and trenches in wafers or substrates. The Bosch process is one of the high-rate DRIE technologies, capable of fabricating 90° vertical walls theoretically [28–30]. The Bosch process alternates between isotropic plasma etching and deposition of a passivation layer, also called pulsed or time-multiplexed etching. The etching–deposition procedures will be repeated until all of the demands are satisfied. However, it is hard to obtain a sidewall precisely vertical to the substrate. Morphology features like inclinations, undulating ripples as well as over-etching are inevitable and critical, which are called the trapezium effect, the ripple effect and footing effect in the following, respectively. The variations induced by these effects manifest roughly as planar sizes (dominated by photolithography and etching processes), planar position offset (dominated by alignment) and vertical sizes (dominated by thickness variations of thin films or the substrate). An ideal beam is a cuboid structure with a length dominated as l, a width as w, and the thickness as h. The cross section was supposed to be a standard rectangle, while, in fact, it appeared as the side-view given in Figure 1a–c. The beams illustrated in Figure 1 are fabricated by DRIE technology, with the sidewall inclination around 84.3°, the undulating ripples about 120°, and the over-etching circled in red of Figure 1c.

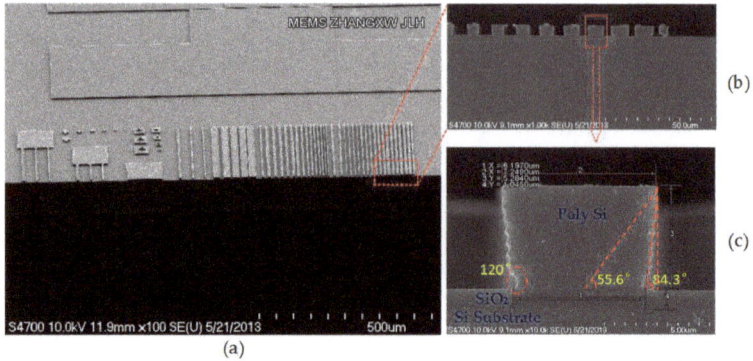

Figure 1. Scanning electron microscope (SEM) cross section for deep reactive ion etching (DRIE) beams: (a) top view of beam array; (b) side-view of the beams labeled in (a); (c) side-view of the single beam labeled in (b), where the footing effect reflects as an arc angle approximating 120°, the inclination angle of 84.3° and 55.6° in the worst-case.

To reveal the significance of manufacturing process variations, a simplified doubly-clamped beam is illustrated in Figure 2. The thickness and length of the beam are assumed to undergo the same manufacturing process variation as 0.05 μm. Thus, for a 200 μm long and 2 μm thick beam, the relative error for the length equals 0.05%, while it is 5.0% for the thickness case. The parameter thickness is obviously more sensitive to process variations. Combined with the usually quadruple relationship of length in a beam's frequency, the relative error diminishes to 0.0006%. This finding reveals that the length variation can be ignored in certain cases to simplify the analysis models.

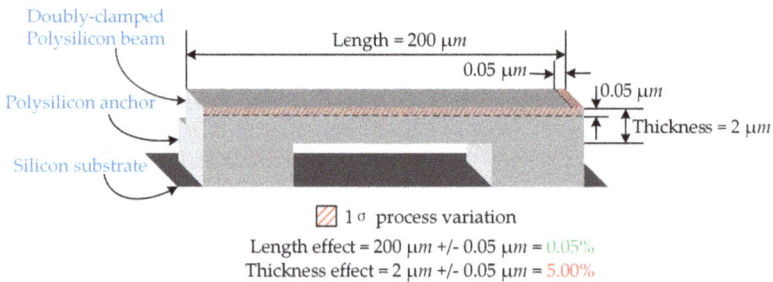

Figure 2. Side-view of a doubly-clamped beam section, assumed with length 200 ± 0.05 μm and thickness 2 ± 0.05 μm. The red part stands for manufacturing process variations, marked as 0.05% and 5.0% on beam length and thickness, respectively.

3. Problem Solution

Studies on MEMS devices have pointed out that manufacturing process variations have a close relationship with performance drift and device failure [22,25,31]. As the basic element in MEMS, the resonant frequency of the doubly-clamped beam underlies the majority of engineering designs. Assuming the section as a plane, the doubly-clamped beam can be treated as an Euler–Bernoulli beam. Without initial buckling, the differential equation for lateral oscillation can be expressed as

$$\overline{EI}\frac{\partial^4 z(x,t)}{\partial x^4} - \overline{\sigma A}\frac{\partial^2 z(x,t)}{\partial x^2} = -\overline{\rho A}\frac{\partial^2 z(x,t)}{\partial t^2}, \tag{1}$$

where $\overline{EI}$ is the bending stiffness, $\overline{\rho A}$ is the linear density, $\overline{\sigma A}$ is the axial load, and $z(x,t)$ is the displacement along the z-axis. Ignoring the residual stress, the resonant frequency of the doubly-clamped beam approximates as [32,33]

$$f_i = \frac{1}{2\pi}(k_i l)^2 \sqrt{\frac{\overline{EI}}{\rho A l^4}},$$ (2)

in which $k_i l$ stands for the coefficient of the ith mode of vibration, and the first three values as $k_1 l = 4.730, k_2 l = 7.853, k_3 l = 10.996$.

3.1. Effect of a Single Factor

Apart from geometrical size errors that can be presented in the behavior equations (refer to Appendixs A and B), morphology changes play an important role in the variability of devices' performance and yield. Appropriate models are needed to reflect the main causes that result from manufacturing uncertainties. As stated in Section 2, process variations are mainly rooted in the trapezium, footing and ripple effect. These effects primarily occur in the Bosch process, Reactive ion etching (RIE) for silicon-on-insulator (SOI) structures, as well as time-multiple-deep deposition (TMDE), respectively. More details on the model building are shown in Appendix A.

The cross section of the beam has been transformed from an ideal rectangle to a trapezoid profile due to process variations. This phenomenon is defined as the trapezium effect, which can be divided into the positive trapezium effect (the trapezoid angle $\theta > 0$) and the negative trapezium effect (the trapezoid angle $\theta < 0$). Figure 3 represents the latter, where Figure 3a denotes the section of a doubly-clamped beam. Figure 3b extracts its cross section models and parameters in the coordinate system. In such a case, the resonant frequency changes into:

$$f_t = \frac{(k_i l)^2}{6\pi} \sqrt{\frac{Eh^2\left(b_1^2 + b_2^2 + 4b_1 b_2\right)}{2\rho l^4 (b_1 + b_2)^2}}.$$ (3)

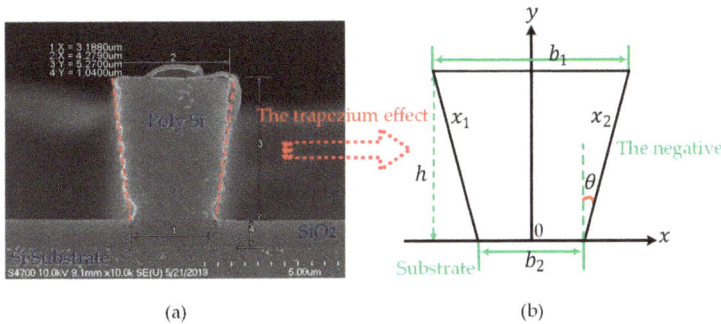

(a)   (b)

Figure 3. The trapezium effect of a doubly-clamped beam: (**a**) SEM side-view, fabricated by the Bosch process and un-released; (**b**) the coordinate sketch-map for the negative.

The footing effect usually shows up in the RIE/DRIE process of SOI structures. It causes inhomogeneous distributions of the mass and stiffness, even making the structure part collapse. The over-etching height h_f and horizontal over-etching width w_f are viewed as the key elements, as demonstrated in Figure 4. Without careful processing control or wide enough width, the structure is prone to crash down, inferred from Figure 4a. The footing effect changes the resonant frequency into:

$$f_f = \frac{(k_i l)^2}{2\pi} \sqrt{\frac{E\left(\frac{1}{3}b_1 h^3 - \frac{1}{6}w_f h_f^3 - \frac{\left(\frac{1}{2}b_1 h^2 - \frac{1}{3}w_f h_f^2\right)^2}{b_1 h - w_f h_f}\right)}{\rho l^4 \left(b_1 h - w_f h_f\right)}}. \tag{4}$$

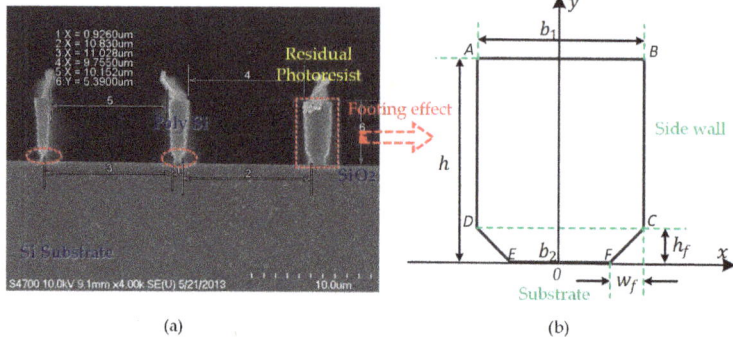

Figure 4. Footing effect of a doubly-clamped beam: (**a**) SEM side-view, fabricated by the Bosch process and un-released; (**b**) the coordinate sketch-map for the structure circled in red in (**a**).

Ripples are presented on the rough side walls of structures with high aspect ratios, owing to the TMDE technology. It is defined as the ripple effect, referring to the simplified model in Figure 5a,b. Given the ripple arc ranging from 30° to 180° and single ripple height from 0.1 μm to 1 μm, simulations in ANSYS have suggested that the deciding element in the ripple effect is single ripple height, rather than ripple arc with an error within 2.0%. Therefore, single ripple height, *t*, can be treated as a key when dealing with the ripple effect.

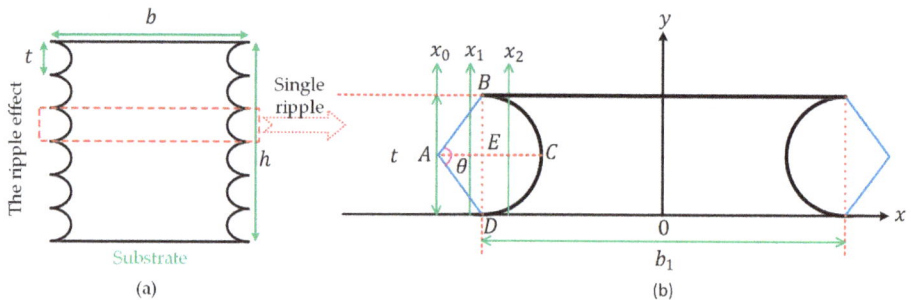

Figure 5. The ripple effect of a doubly-clamped beam: (**a**) a schematic diagram of the model; (**b**) the coordinate sketch-map of single ripple as one unit.

In such a case, the resonant frequency changes into:

$$f_r = \frac{(k_i l)^2}{2\pi} \sqrt{\frac{E\left(-\frac{3\sqrt{3}+8\pi}{144}ht^3 - \frac{2\pi-3\sqrt{3}}{72}h^3 t + \frac{1}{12}b_1 h^3\right)}{\rho l^4 \left(b_1 h - \frac{2\pi-3\sqrt{3}}{6}ht\right)}}, \tag{5}$$

where the meaning of the symbols is marked in Figure 5.

3.2. Effect of Multiple Factors

Sensitivity analyses on the effects mentioned above are conducted, shown in Figure 6. With accurate processing control and mature techniques, single arc height can be restricted to be less than 0.01 µm so that the ripple effect can be diminished, as shown in Figure 6a. Under these circumstances, models can be simplified into two critical effects: the trapezium effect and footing effect. Thus, when side walls are thought to be smooth, the resonant frequency of doubly-clamped beams can be expressed as:

$$f_{tf} = \frac{(k_i l)^2}{2\pi} \sqrt{\frac{E\left(\frac{h^3}{36} \cdot \frac{b_1^2 + b_2^2 + 4b_1 b_2}{b_1 + b_2} - 2\left(\frac{w_f h_f^3}{36} + \frac{w_f h_f}{2}\left(\frac{h_f}{3} - \frac{h(2b_1 + b_2)}{3(b_1 + b_2)}\right)^2\right)\right)}{\rho l^4 \left(\frac{b_1 + b_2}{2} h - w_f h_f\right)}}. \tag{6}$$

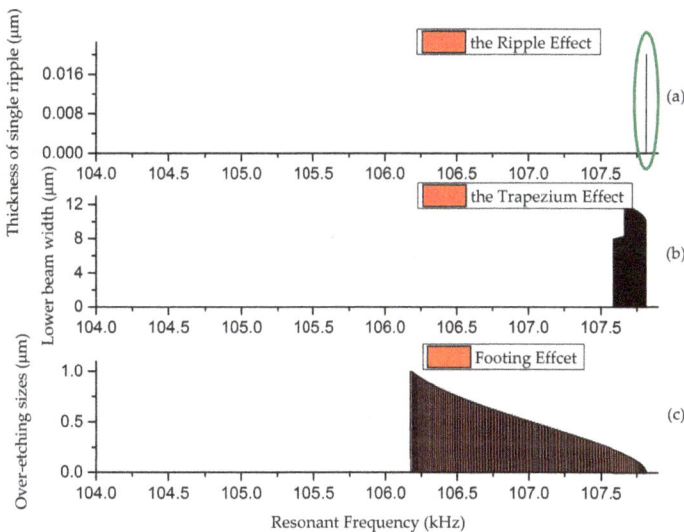

Figure 6. Sensitivity analysis on the effects influencing the resonant frequency of doubly-clamped beams. (**a**) is the result with considering the ripple effect. The narrow band circled by green equals the working part. (**b**,**c**) are cases for the trapezium effect and footing effect, where the cross section is treated as isosceles trapezoid in (**b**) and the over-etching sizes longitudinally and laterally are equal in (**c**).

However, side walls cannot be treated as smooth all the time. Models containing the three effects simultaneously are essential. The corresponding coordinate is illustrated as Figure 7, where $b_1 = b + 2h\tan\theta$, $b_2 = b_1 - 2w_f$. When the gap between beams is wide enough, like 6 µm or wider, the cross section can be assumed to be continuous, repeated and symmetrical. Assumptions can be raised that the central axis of the cross section equals the central axis of the trapezoid $ABCD$, which means $y = y_c$, and that the ripple is a semicircle. The moment of inertia, relative to central axis $y = y_1$, equals the condition of axis z_1, approximately. Axis z_1 is in the direction of the semicircle radius and perpendicular to the side waist of trapezoid $ABCD$, when the semicircle vibrates longitudinally.

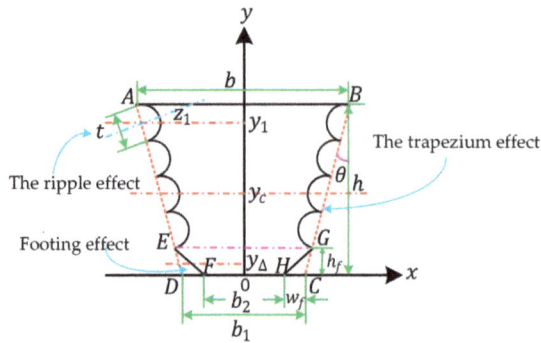

Figure 7. The coordinate sketch-map for the cross section of the doubly-clamped beam considering the three effects simultaneously (only four ripples for illustration).

The above results suggest that the resonant frequency changes into (refer to Appendix B for more details on model building):

$$f_{tfr} = \frac{(k_i l)^2}{2\pi} \sqrt{\frac{E\left(\frac{h^3}{36} \cdot \frac{b^2 + b_1^2 + 4bb_1}{b + b_1} - 2\left(\frac{w_f h_f^3}{36} + \frac{A_f}{2}\left(\frac{h_f}{3} - \frac{2b + b_1}{3(b_1 + b)}h\right)^2\right) - 2\sum_{i=1}^{N}\left(\frac{1}{2} \cdot \frac{1}{64}\pi r^4 + \frac{\pi r^2}{2}(y_1 - i*2r\cos\theta - y_c)^2\right)\right)}{\rho l^4\left(\frac{(b_1 + b)h}{2} - w_f h_f - N\pi r^2\right)}} \tag{7}$$

4. Analysis and Results

Geometric features of MEMS devices usually do not comply with the design value, with the typical error around 5.0% [34,35] during the manufacturing processes. Design parameters for the doubly-clamped beam are listed in Table 1.

Table 1. Design parameters for a doubly-clamped beam.

Structure Parameters	Values
Beam length l/μm	200
Beam width b/μm	4
Beam thickness h/μm	2
Young's modulus $E/$GPa	158
Material density $\rho/$kg/μm^3	2.23×10^{-15}

Multi-field models are always complicated for the complex mechanism of MEMS devices. The situation deteriorates with stochastic manufacturing uncertainties. Leaving the cost alone, with repeated adjustment or tape-out of test structures, only a small percent of the data is acceptable. Due to the lack of manufacturing data, the trial-and-error method is not optimal. Adequate models underlying process variations should be developed.

Simulations are conducted in ANSYS WORKBENCH 14.5 (ANSYS, Pittsburgh, PA, USA), with the solver SOLID45 (ANSYS, Pittsburgh, PA, USA) [27]. The mesh method uses tetrahedrons with patch conforming. Lateral vibrations are considered to perform model verifications, as shown in Figure 8. The number of ripples is supposed to be 5 and 10, in order to simplify the validations. Assuming small variation ranges in key elements of the trapezium and footing effect, analyses of curves in Figure 8 are carried out. The resonant frequency is found to be in direct proportion to the number of ripples N, with improvement in stable behavior along with larger N. Morever, the frequency manifests a reverse proportion to the angle, undergoing serious shifts in the wake of deteriorative footing effect.

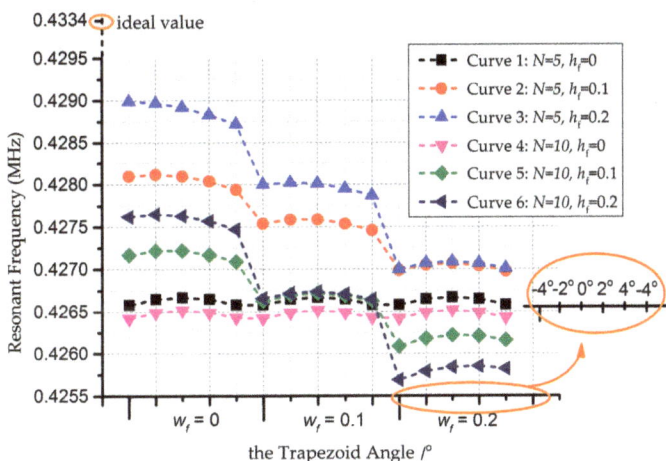

Figure 8. Change curves of the resonance frequency for the modified model of the doubly-clamped beam. The *x* coordinate is divided into three intervals, each of which is denoted as magnified in the orange circle.

In light of the above development, comparisons between modified models and ANSYS are conducted, with consideration of single effects, respectively. Figure 9a describes errors under the trapezium effect, with the bias less than 2.5%. It turns out that the model and simulations share a similar trend. Figure 9b states the situation for the ripple effect. As the situation for a high value of the arc height rarely occurs during processing, the result with an error of 10.1% is not accurate or applicable for further study. The first two results are receivable in general, limiting the errors within 2.0%. Confining errors within 3%, over-etching longitudinally introduces more variabilities in resonant frequency, referred to in Figure 9c. Complicated structures will result in larger differences. Analyses are conducted in Figure 9d to explain the footing and trapezium effects. In pursuit of less variabilities in frequency, the negative trapezium effect is proved to be effective. In addition, the footing effect occupies the dominant position, compared with the trapezium effect, according to Figure 6b,c.

Furthermore, comparisons have been established between modified models and FE analysis. The number of ripples N is assigned to 10, while $w_f = h_f = 0.2$ µm in the footing effect. The curves share an error within 2.6%, according to Figure 10. The outliers in the upper right corner of Figure 10 suggest biases of ANSYS simulations. This occurrence is attributed to the unpractical assumptions that angles in the positive trapezium effect can be 20°, in which case the principles of Timoshenke beams cannot be applied directly. However, the assumptions give credit to the negative case for the existence of the footing effect. The two curves trend similarly in general, which confirms the acceptability of the modified models.

Direct Monte Carlo (MC) simulations are performed based on the modified models. Yield is defined as a factor of the proportion falling into the same distribution range. Doubly-clamped beams with 400 µm length, 10 µm width and 4 µm thickness are raised as an example in these simulations. Hypotheses are proposed that all the parameters concerned comply with the Gaussian's distribution, along with the same process variation ±0.5 µm. The sampling numbers for MC simulations range from 100,000 to 1,000,000. The relative error for frequency turns out to be around 7.9% and an angle of around ±7° when considering the trapezium effect. The resonant frequency reduces from 22.7% to 33.0% while the yield decreases to nearly 67.0% under the footing effect. When doubling the numbers of ripples, relative errors for the resonant frequency can be improved about 1.0%.

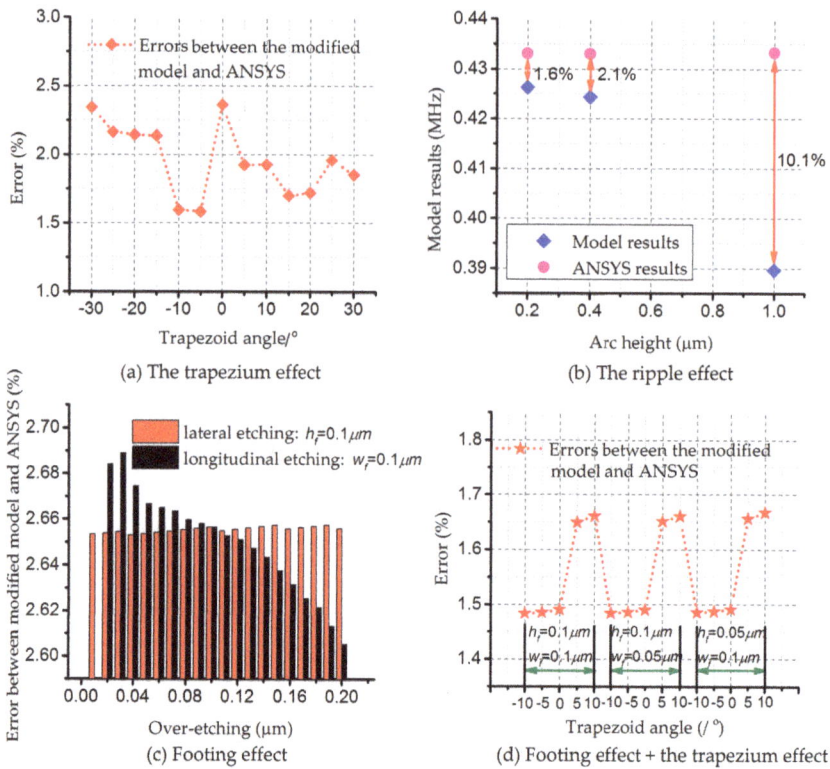

Figure 9. Comparisons between modified models and ANSYS while considering models related to Equations (3)–(6) (the errors along the vertical axis defined as: Error = |model results—ANSYS results|/model results): (**a**) the trapezium effect; (**b**) the ripple effect; (**c**) the footing effect, the red for lateral over-etching and black for over-etching longitudinally; and (**d**) the footing effect and the trapezium effect.

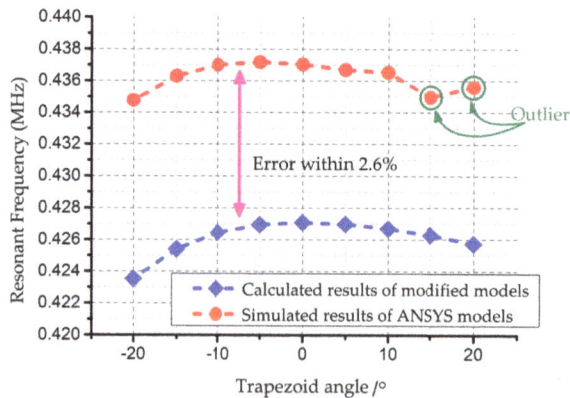

Figure 10. Change curves of the resonance frequency for the doubly-clamped beam. Process variations are set as ±0.5 μm. The dots in green circles are outliers of the simulated results.

5. Conclusions

This paper has considered the problems existing in the Bosch process and their negative influences on MEMS doubly-clamped beam performance. Modified models of doubly-clamped beams were built, with consideration for the trapezium, footing, and ripple effects respectively and simultaneously. The relative performance error was restricted to 10.0%, with a yield of about 77.3% if process variations were assumed to be in the same range. FE verifications have been performed to validate the models built in this study, indicating that the heavy simulation work can be substituted in some cases by applying the models.

The model results can be viewed as the guidance for design cycle optimizations. Designers can directly figure out the key elements in the etching process by reconsidering the design sizes and shapes, and eventually compensate the errors brought by process variations to improve the yield.

Other critical elements such as residual stress, gaps between beams, or variations in Young's modulus were not considered in this paper and will be discussed in future work. Moreover, diversified distribution forms such as quasi-Gaussian can be applied in MC methods and will be the focus of future research.

Acknowledgments: The authors would like to acknowledge the support offered by the National High Technology Development Program of China under Contract 2015AA042604.

Author Contributions: Lili Gao developed the model, and Lili Gao and Zai-Fa Zhou jointly performed the simulations and the data analysis. Qing-An Huang proposed the idea for the model and edited this manuscript.

Conflicts of Interest: The authors declare no conflict of interest.

Appendix A. Single Factor Effect

Appendix A.1. The Trapezium Effect

During the DRIE processes, the ideal rectangle profile of a beam can be transferred to a trapezoid profile due to process variations. Usually, the cases are divided into the positive trapezium effect (the trapezoid angle $\theta > 0$) and the negative trapezium effect (the trapezoid angle $\theta < 0$), shown in Figure 3. For the ideal doubly-clamped beam, the resonance frequency can be expressed as:

$$f = \frac{(k_i l)^2}{4\pi}\sqrt{\frac{Ew^2}{3\rho l^4}}. \tag{A1}$$

Inducing x-y coordinates (the same way for the positive), the upper width b_1 is:

$$b_1 = b_2 - 2h \cdot \tan\theta. \tag{A2}$$

The neutral axis could be denoted by y_{na}, along with the linear equations of x_1 and x_2, that is:

$$y_{na} = \frac{S_z}{A_t} = \frac{2b_1 + b_2}{3(b_1 + b_2)}h. \tag{A3}$$

Here, S_z is the static moment, and A_t is the current cross-sectional area.
Then, the moment of inertia is expressed as:

$$I_t = \int_A y^2 dA == \frac{h^3\left(b_1^2 + b_2^2 + 4b_1 b_2\right)}{36(b_1 + b_2)}. \tag{A4}$$

Thus, the resonant frequency of the doubly-clamped beam for the modified model considering the trapezoid effect changes into:

$$f_t = \frac{(k_i l)^2}{6\pi} \sqrt{\frac{Eh^2\left(b_1^2 + b_2^2 + 4b_1 b_2\right)}{2\rho l^4 (b_1 + b_2)^2}}. \tag{A5}$$

Appendix A.2. The Footing Effect

The critical element in the footing effect is the over-etching height h_f and horizontal over-etching width w_f, as shown in Figure 4. According to the properties for the moment of inertia, the moment of inertia relative to the neutral axis for the cross section $ABCFED$ can be expressed as:

$$I'_{ABCFED} = I'_{ABCD} + I'_{CDEF}, \tag{A6}$$

where

$$I'_{ABCD} = \frac{1}{12}b_1\left(h - h_f\right)^3, \tag{A7}$$

$$I'_{CDEF} = \frac{h_f^3}{36\left(b_1 - w_f\right)}(3b_1^2 - 6b_1 w_f + 2w_f^2). \tag{A8}$$

Then, regards to the parallel-axis theorem, the moment of inertia relative to the neutral axis y_c for the cross section $ABCFED$ becomes

$$I_f = I_{ABCFED} = I_{ABCD} + I_{CDEF} = I'_{ABCD} + b_1\left(h - h_f\right)(y_c - y_1)^2 + I'_{CDEF} + h_f\left(b_1 - w_f\right)(y_c - y_2)^2. \tag{A9}$$

Here, y_1 and y_2 denote the neutral axis of $ABCD$ and $CDEF$, relative to each own centroid, respectively, and

$$y_1 = \frac{h + h_f}{2}, \; y_2 = \frac{3b_1 - 2w_f}{26b_1 - 6w_f}h_f, \; y_c = \frac{\frac{1}{2}b_1 h^2 - \frac{1}{3}w_f h_f^2}{b_1 h - w_f h_f}. \tag{A10}$$

Thus, the frequency of the doubly-clamped beam for the modified model considering the footing effect changes into:

$$f_f = \frac{(k_i l)^2}{2\pi} \sqrt{\frac{E\left(\frac{1}{3}b_1 h^3 - \frac{1}{6}w_f h_f^3 - \frac{\left(\frac{1}{2}b_1 h^2 - \frac{1}{3}w_f h_f^2\right)^2}{b_1 h - w_f h_f}\right)}{\rho l^4\left(b_1 h - w_f h_f\right)}}. \tag{A11}$$

Appendix A.3. The Ripple Effect

The key element of the ripple effect is the height of the ripple, almost nothing to do with the arc of the ripple. Therefore, the height of the ripple, t, is taken as an impact factor of the ripple effect shown as Figure 5. The impacts on frequency mainly arise from t, and the central angle is chosen as $\theta = 60°$, according to the statistical results of tapeout (for the further model optimization, θ could be treated as an independent variable). Similar to the knowledge of moment of inetia, the cross section regarding $x = 0$ under the ripple effect is the sum of each unit, depicted in Figure 5. Symetrically, the moment of inertia of each unit stays the same, regarding the axis of $x = 0$, denoted as I_0. Here, we have $AB = AD = BD = r = t$, $AE = \frac{\sqrt{3}}{2}r$, $CE = r - \frac{\sqrt{3}}{2}r$.

With a series of calculations, the moment of inertia relative to the neutral axis itself for one unit is:

$$I_0 = \frac{b_1 t^3}{12} + \frac{\sqrt{3} - 4\pi}{48}t^4. \tag{A12}$$

Similarly, the sum of the moment of inertia under the ripple effect is expressed as:

$$I_r = \sum I_i = \sum_{i=1}^{i=N}\left(I_0 + \left(\frac{h}{2} - \frac{2i-1}{2}t\right)^2\left(b_1 t - \frac{2\pi - 3\sqrt{3}}{6}t^2\right)\right) = -\frac{3\sqrt{3}+8\pi}{144}ht^3 - \frac{2\pi - 3\sqrt{3}}{72}h^3 t + \frac{1}{12}b_1 h^3. \tag{A13}$$

Finally, the frequency of the doubly-clamped beam for the modified model considering the ripple effect changes into:

$$f_r = \frac{(k_i l)^2}{2\pi} \sqrt{\frac{E\left(-\frac{3\sqrt{3}+8\pi}{144} h t^3 - \frac{2\pi - 3\sqrt{3}}{72} h^3 t + \frac{1}{12} b_1 h^3\right)}{\rho l^4 \left(b_1 h - \frac{2\pi - 3\sqrt{3}}{6} h t\right)}}. \tag{A14}$$

Appendix B. Multiple Factors Effect

Appendix B.1. The Ripple Effect and Footing Effect

As the moment of inertia is essential for model building (the same sequence as it in Appendix A1), the moment of inertia can de deduced from the trapezium effect. If the over-etching part arising from the footing effect is neglected compared with the part in the trapezium effect, the transversal change of the neutral axis due to the footing effect could also be ignored. Applying the corresponding coordinate in Figures 6 and A1, the moment of inertia for the cross section is written as:

$$I_{tf} = I_{ABCD} - I_{DEG} - I_{CFH} = \frac{h^3}{36} \cdot \frac{b_1^2 + b_2^2 + 4b_1 b_2}{b_1 + b_2} - 2\left(\frac{w_f h_f^3}{36} + \frac{w_f h_f}{2}\left(\frac{h_f}{3} - \frac{h(2b_1 + b_2)}{3(b_1 + b_2)}\right)^2\right), \tag{B1}$$

where $s_{tf} = \frac{b_1 + b_2}{2} h - w_f h_f$. The frequency of the doubly-clamped beam for the modified model considering the footing effect and the trapezium effect changes into:

$$f_{tf} = \frac{(k_i l)^2}{2\pi} \sqrt{\frac{E\left(\frac{h^3}{36} \cdot \frac{b_1^2 + b_2^2 + 4b_1 b_2}{b_1 + b_2} - 2\left(\frac{w_f h_f^3}{36} + \frac{w_f h_f}{2}\left(\frac{h_f}{3} - \frac{h(2b_1 + b_2)}{3(b_1 + b_2)}\right)^2\right)\right)}{\rho l^4 \left(\frac{b_1 + b_2}{2} h - w_f h_f\right)}}. \tag{B2}$$

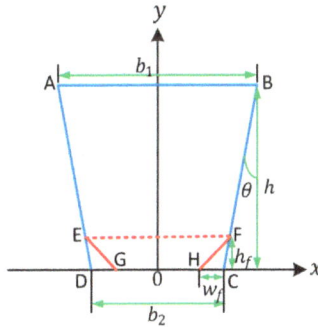

Figure A1. The coordinate sketch-map for the cross section of a doubly-clamped beam while considering the footing effect and the trapezium effect.

Appendix B.2. The Ripple Effect, the Footing Effect, and the Ripple Effect

The corresponding coordinate for the three effects is shown in Figure 6, where $b_1 = b + 2h \tan \theta$, and $b_2 = b_1 - 2w_f$. With the assumptions proposed above, the area and moment of inertia for the cross section are written as:

$$A = \frac{(b_1 + b)h}{2} - w_f h_f - N\pi r^2, \tag{B3}$$

$$I = I_{ABCD} - I_{DEF} - I_{CGH} - \sum I_{arc}. \tag{B4}$$

In addition, it is easy to get:

$$I_{ABCD} = \frac{h^3}{36} \cdot \frac{b^2 + b_1^2 + 4bb_1}{b + b_1},$$ (B5)

$$I_f = I_{DEF} + I_{CGH} = 2\left(\frac{w_f h_f^3}{36} + \frac{A_f}{2}\left(\frac{h_f}{3} - \frac{2b + b_1}{3(b_1 + b)}h \right)^2 \right).$$ (B6)

Models should be developed to analyze arc coordinates for the solution of I_{arc} when the trapezoid angle $\theta > 0$ or $\theta < 0$. The key to the arc coordinates is the position of the first arc (see Figure A2).

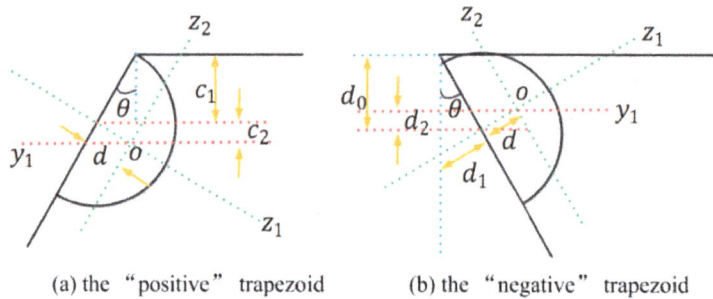

(a) the "positive" trapezoid   (b) the "negative" trapezoid

Figure A2. Coordinate sketch-map for the first arc. (**a**) the positive trapezoid (the upper length lower than the lower length) while the trapezoid angle $\theta > 0$; (**b**) the negative one and $\theta < 0$.

In Figure A2a, z_1 and z_2 are the neutral axes of the semicircle itself, by the point of their intersection o, and y_1 is parallel to y_c. Consequently, the coordinate of y_1 is $y_1 = h - \frac{r}{\cos\theta} - d\tan\theta$. For the condition in Figure A2b, $y_1 = h - \frac{r}{\cos\theta} + (d + r\tan\theta)\sin\theta$. With the assumptions above, the moment of inertia for the semicircle relative to neutral axis z_1 equals $I_0 = \frac{1}{2} \cdot \frac{1}{64}\pi r^4$, and then the moment of inertia relative to the cross section for the ith arc becomes

$$I_i = \frac{1}{2} \cdot \frac{1}{64}\pi r^4 + \frac{\pi r^2}{2}(y_1 - i * 2r\cos\theta - y_c)^2.$$ (B7)

Therefore,

$$I = I_{ABCD} - I_f - 2\sum_{i=1}^{n} I_i.$$ (B8)

Then, the frequency of the doubly-clamped beam for the modified model considering three main effects changes into

$$f_{tfr} = \frac{(k_i l)^2}{2\pi} \sqrt{\frac{E\left(\frac{h^3}{36} \cdot \frac{b^2 + b_1^2 + 4bb_1}{b + b_1} - 2\left(\frac{w_f h_f^3}{36} + \frac{A_f}{2}\left(\frac{h_f}{3} - \frac{2b + b_1}{3(b_1 + b)}h \right)^2 \right) - 2\sum_{i=1}^{N}\left(\frac{1}{2} \cdot \frac{1}{64}\pi r^4 + \frac{\pi r^2}{2}(y_1 - i*2r\cos\theta - y_c)^2 \right) \right)}{\rho l^4\left(\frac{(b_1 + b)h}{2} - w_f h_f - N\pi r^2 \right)}}.$$ (B9)

References

1. Reza Ghodssi, P.L. *MEMS Materials and Processes Handbook*; Springer: New York, NY, USA, 2011; pp. 1067–1076.
2. Oropeza-Ramos, L.A.; Burgner, C.B.; Turner, K.L. Robust micro-rate sensor actuated by parametric resonance. *Sens. Actuators A Phys.* **2009**, *152*, 80–87. [CrossRef]
3. Shavezipur, M.; Ponnambalam, K.; Khajepour, A.; Hashemi, S.M. Fabrication uncertainties and yield optimization in MEMS tunable capacitors. *Sens. Actuators A Phys.* **2008**, *147*, 613–622. [CrossRef]

4. Lee, M.C.; Kang, S.J.; Jung, K.D.; Choa, S.-H.; Cho, Y.C. A high yield rate MEMS gyroscope with a packaged siog process. *J. Micromech. Microeng.* **2005**, *15*, 2003. [CrossRef]

5. Yeom, J.; Wu, Y.; Selby, J.C.; Shannon, M.A. Maximum achievable aspect ratio in deep reactive ion etching of silicon due to aspect ratio dependent transport and the microloading effect. *J. Vac. Sci. Technol. B* **2005**, *23*, 2319–2329. [CrossRef]

6. Johnson, B.N.; Mutharasan, R. Biosensing using dynamic-mode cantilever sensors: A review. *Biosens. Bioelectron.* **2012**, *32*, 1–18. [CrossRef] [PubMed]

7. Lang, W. Silicon microstructuring technology. *Mater. Sci. Eng. R Rep.* **1996**, *17*, 1–55. [CrossRef]

8. Younis, M.I. *MEMS Linear and Nonlinear Statics and Dynamics*; Springer: New York, NY, USA, 2010; p. 453.

9. Milor, L. A survey of yield modeling and yield enhancement methods. *IEEE Trans. Semicond. Manuf.* **2013**, *26*, 196–213. [CrossRef]

10. Islam, M.F.; Ali, M.A.M.; Majlis, B.Y. Probabilistic analysis of micro-machined fixed-fixed beam for reliability. *IET Micro Nano Lett.* **2008**, *3*, 95–100. [CrossRef]

11. Liu, M.; Maute, K.; Frangopol, D.M. Multi-objective design optimization of electrostatically actuated microbeam resonators with and without parameter uncertainty. *Reliab. Eng. Syst. Saf.* **2007**, *92*, 1333–1343. [CrossRef]

12. Rong, L.; Paden, B.; Turner, K. MEMS resonators that are robust to process-induced feature width variations. *J. Microelectromech. Syst.* **2002**, *11*, 505–511. [CrossRef]

13. Mawardi, A.; Pitchumani, R. Design of microresonators under uncertainty. *J. Microelectromech. Syst.* **2005**, *14*, 63–69. [CrossRef]

14. Bagherinia, M.; Bruggi, M.; Corigliano, A.; Mariani, S.; Lasalandra, E. Geometry optimization of a lorentz force, resonating MEMS magnetometer. *Microelectron. Reliab.* **2014**, *54*, 1192–1199. [CrossRef]

15. Luo, Z.; Wang, X.; Jin, M.; Liu, S. MEMS gyroscope yield simulation based on Monte Carlo method. In Proceedings of the 2012 IEEE 62nd Electronic Components and Technology Conference, 29 May–1 June 2012; pp. 1636–1639.

16. Mirzazadeh, R.; Eftekhar Azam, S.; Mariani, S. Micromechanical characterization of polysilicon films through on-chip tests. *Sensors* **2016**, *16*, 1191. [CrossRef] [PubMed]

17. Mirzazadeh, R.G.A.; Mariani, S. Assessment of overetch and polysilicon film properties through on-chip tests. In Proceedings of the 2nd International Electronic Conference on Sensors and Applications, Basel, Switzerland, 15–30 November 2015.

18. Mirzazadeh, R.M.S. Assessment of micromechanically-induced uncertainties in the electromechanical response of MEMS devices. In Proceedings of the 3rd International Electronic Conference on Sensors and Applications, Basel, Switzerland, 15–30 November 2016.

19. Shavezipur, M.; Ponnambalam, K.; Hashemi, S.M.; Khajepour, A. A probabilistic design optimization for MEMS tunable capacitors. *Microelectron. J.* **2008**, *39*, 1528–1533. [CrossRef]

20. Allen, M.; Raulli, M.; Maute, K.; Frangopol, D.M. Reliability-based analysis and design optimization of electrostatically actuated MEMS. *Comput. Struct.* **2004**, *82*, 1007–1020. [CrossRef]

21. Pfingsten, T.; Herrmann, D.J.L.; Rasmussen, C.E. Model-based design analysis and yield optimization. *IEEE Trans. Semicond. Manuf.* **2006**, *19*, 475–486. [CrossRef]

22. Vudathu, S.P.; Duganapalli, K.K.; Laur, R.; Kubalinska, D.; Bunse-Gerstner, A. Yield analysis via induction of process statistics into the design of MEMS and other microsystems. *Microsyst. Technol.* **2007**, *13*, 1545–1551. [CrossRef]

23. Engesser, M.; Buhmann, A.; Franke, A.R.; Korvink, J.G. Efficient reliability-based design optimization for microelectromechanical systems. *IEEE Sens. J.* **2010**, *10*, 1383–1390. [CrossRef]

24. Agarwal, N.; Aluru, N.R. Stochastic modeling of coupled electromechanical interaction for uncertainty quantification in electrostatically actuated MEMS. *Comput. Methods Appl. Mech. Eng.* **2008**, *197*, 3456–3471. [CrossRef]

25. Dewey, A.; Ren, H.; Zhang, T. Behavioral modeling of microelectromechanical systems (MEMS) with statistical performance-variability reduction and sensitivity analysis. *IEEE Trans. Circuits Syst. II Analog Digit. Signal Process.* **2000**, *47*, 105–113. [CrossRef]

26. Vudathu, S.P.; Laur, R. A design methodology for the yield enhancement of MEMS designs with respect to process induced variations. In Proceedings of the 57th Electronic Components and Technology Conference, Sparks, NV, USA, 29 May–1 June 2007.

27. ANSYS Academic Research, Release 16.2, Help System, ANSYS, Inc. Available online: http://ansys.com/products/academic (accessed on 3 March 2017).
28. Laermer, F.; Schilp, A. Method of Anisotropically Etching Silicon. U.S. Patent US5501893, 26 March 1996.
29. Laermer, F.; Schilp, A. Method for Anisotropic Etching of Silicon. U.S. Patent US6284148, 4 September 2001.
30. Laermer, F.; Schilp, A. Method of Anisotropic Etching of Silicon. U.S. Patent US6531068, 11 March 2003.
31. Wu, B. A statistically optimal macromodeling framework with application in process variation analysis of MEMS devices. In Proceedings of the 2012 IEEE 10th International New Circuits and Systems Conference (NEWCAS), Montreal, QC, Canada, 17–20 June 2012; pp. 221–224.
32. Liu, C. *Foundations of MEMS*; Prentice Hall Press: Upper Saddle River, NJ, USA, 2011; p. 560.
33. Weaver, W., Jr.; Timoshenko, S.P.; Young, D.H. *Vibration Problems in Engineering*; Wiley: New York, NY, USA, 1974; pp. 415–455.
34. Cowen, A.; Hardy, B.; Mahadevan, R.; Wilcenski, S. *PolyMUMPs Design Handbook*; MEMSCAP Inc.: Durham, NC, USA, 2011.
35. Cowen, A.; Hames, G.; Monk, D.; Wilcenski, S.; Hardy, B. *SoiMUMPs Design Handbook*; MEMSCAP Inc.: Durham, NC, USA, 2011.

micromachines

MDPI

Article

Low Temperature Plasma Nitriding of Inner Surfaces in Stainless Steel Mini-/Micro-Pipes and Nozzles

Tatsuhiko Aizawa [1],* and Kenji Wasa [2]

[1] Department of Engineering and Design, Shibaura Institute of Technology, 3-19-10 Shibaura, Minato-City, Tokyo 108-8548, Japan

[2] Research and Development Division, TECDIA, Co., Ltd., 4-3-4 Shibaura, Minato-City, Tokyo 108-0023, Japan; k_wasa@tecdia.co.jp

* Correspondence: taizawa@sic.shibaura-it.ac.jp; Tel.: +81-3-6722-2741

Academic Editors: Hans Nørgaard Hansen and Guido Tosello
Received: 9 January 2017; Accepted: 11 May 2017; Published: 13 May 2017

Abstract: Metallic miniature products have been highlighted as mini-/micro-structural components working as a precise mechanism, in dispensing systems, and in medical operations. In particular, the essential mechanical parts such as pipes and nozzles have strength and hardness sufficient for ejecting viscous liquids, solders, and particles. A low-temperature plasma nitriding process was proposed as a surface treatment to improve the engineering durability of stainless steel mini-/micro-pipes and nozzles. Various analyses were performed to describe the inner nitriding process only, from the inner surface of pipes and nozzles to their depth in thickness. AISI316 pipes and AISI316/AISI304 nozzle specimens were used to demonstrate by plasma nitriding for 14.4 ks at 693 K that their inner surfaces had a hardness higher than 800 HV.

Keywords: micro-fabrication; mini- and micro-nozzles; stainless steels; plasma nitriding; inner surfaces; hollow cathode device; nitrogen interstitials; hardness

1. Introduction

Pipes and nozzles are typical mechanical parts for transporting gaseous and liquid media and depositing liquid- or melt-drops onto material surfaces and interfaces in joining, soldering, and drawing processes [1]. In cellular phones, many devices and sensors are joined onto liquid crystal panels by adhesives, which are deposited onto the joined interface [2]. For example, digital camera units are integrated into cellular phones by accurately joining a series of functional parts. Dimensional accuracy and integrity is preserved by joining via adhesive droplets. These nozzle outlets contact and hit the joined surfaces; the stainless steel nozzles often deform and become damaged in the joining process. The ruby nozzle is often selected for sufficient hardness in operation.

High-density plasma nitriding has been used for hardening metals and alloys such as stainless steels, tool steels, and titanium and aluminum alloys with the use of DC-biased metallic tubes (referred to here as hollow-cathode devices) [3–6]. In these processes, these metallic alloy parts and components are efficiently plasma-nitrided in hollow-cathode devices. For example, AISI-420 martensitic stainless steel die-parts have been plasma-nitrided at 673 K for 14.4 ks to have a hardness higher than 1000 HV [3,7,8]. In addition, AISI304 and AISI316 stainless steel plates were also plasma-nitrided at 673 K for 14.4 ks by 70 Pa to have a nitrided layer thickness higher than 80 μm and a hardness higher than 1400 HV, respectively [9–11].

In the present study, this surface treatment, with the use of a hollow-cathode device, was employed to harden the inner surfaces of mini-/micro-nozzles and pipes for the joining process. The inner diameter of their through-holes ranged from 50 μm to 10 mm. The hydrogen–nitrogen mixed gas was blown into the inlet of these nozzles. The nitrided through-hole surfaces were observed by Scanning

Electron Microscopy (SEM), and their hardness was measured via micro-Vickers testing. In the case of the micro-nozzles with a through-hole diameter of 50 μm, their inner surface hardness increased up to 950 HV on average after plasma nitriding at 693 K for 14.4 ks.

2. Experimental Procedure

2.1. RF-DC Plasma Generation System

The low-temperature plasma nitriding process with use of the RF-DC plasmas had several superior features to the conventional processes [12,13]. The lower holding temperature with shorter duration time was cost-effective, and, the nitrided surface was free from roughing with less nitride precipitates. Since the inner surface roughening was disliked for nozzles, the present nitriding process with nitrogen super-saturation can afford to harden without significant roughing as suggested by [14]. Figure 1a illustrated the RF-DC plasma nitriding system. The dipole electrodes worked to ignite the RF-plasma, which was attracted to the DC-biased plate. Since the vacuum chamber was electrically neutral, the above RF-field as well as DC-biased field were arbitrarily placed and controlled independently in the chamber. This concept of plasma generation and control was put into practice. In Figure 1b, this RF-DC plasma nitriding system mainly consisted of six sectors: (1) vacuum chamber; (2) RF-generator; (3) control panel; (4) power generator; (5) evacuation units; and, (6) carrier gas supply.

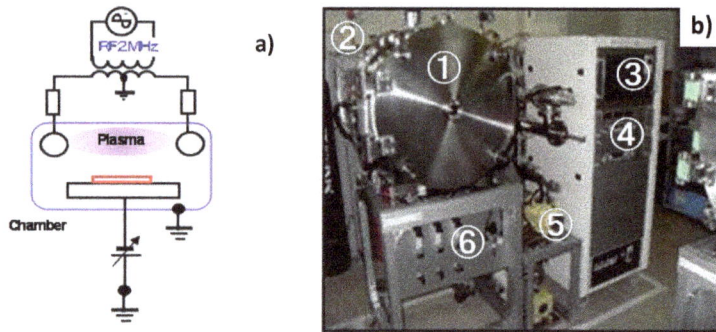

Figure 1. RF-DC plasma process system. (**a**) Illustration of system. (**b**) Developed plasma nitriding system in this study.

2.2. Hollow Cathode Device

In order to ignite the nitrogen plasmas only in the nozzle through-holes as well as the pipes, the RF-plasmas were confined in the inside of electrically conductive space. The hollow cathode effect to increase the ion and electron densities in the hollow helped to design how to accommodate the high-density nitrogen plasmas selectively in the through hole. Figure 2a illustrated a typical hollow cathode setup. The mixed carrier gas of nitrogen and hydrogen with the specified flow rate ratio was blown into the inlet of hollow tube through the flexible metal-bellows.

The hollow tube was DC-biased so that the nitrogen/hydrogen RF plasma was confined in this through-hole; the ionization of nitrogen and hydrogen gases should take place only in its inside. In the preliminary studies [15], {N^*, N^+, NH (or NHx)} were detected by Emissive-light Optical Spectroscopy (EOS) and used to optimize the nitriding conditions. After three dimensional electromagnetic analyses in [16], the electrons are confined in the hollow by the electromagnetic shield at the outlet to enhance the ionization process. As shown in Figure 2b, the emissive light from the inside of hollow was intensified in the experiments with the RF voltage of 250 V and the DC-bias of −500 V. Figure 3 depicted a typical ion density distribution, measured by the Langmuir probe from the inlet toward the outlet of hollow with the length of 120 mm [16]. Since the ionization was enhanced by the carrier gas flow, the nitrogen ion density increased monotonically toward the outlet of hollow. This high density abruptly decreased down in one or two orders just at the outside of this hollow. This ion density distribution was preferable to make selective nitriding of the through holes; in particular, the inner surface of nozzle outlet could be efficiently nitrided. After the Langmuir probe measurement and emissive-light optical spectroscopy, the species of activated nitrogen atoms and NHx-radicals might be responsible for nitriding process of the austenitic stainless steel pipes and nozzles as discussed in [17]. As suggested in [18], the flow rate ratio of H$_2$ to N$_2$ in the mixture gas also affected the NH-radical yields in the generated plasmas. The nitriding condition with the RF voltage of 250 V, the DC-bias of −500 V, the pressure of 70 Pa and the N$_2$/H$_2$ ratio by 160 mL/min to 30 mL/min was used as standard processing parameters in the following experiments. The holding temperature was measured by the thermocouple, which was embedded in the jig to fix the pipe and nozzle on the cathode plate. The heating unit was on/off controlled to keep the measured temperature by 693 K.

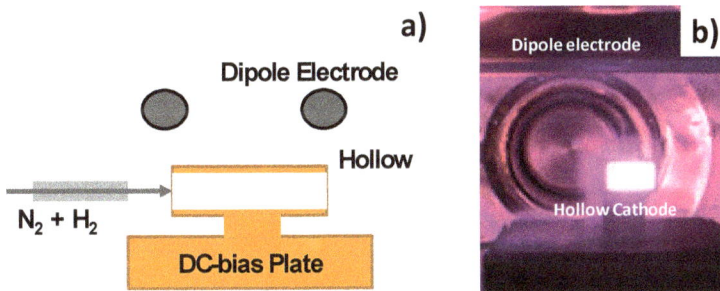

Figure 2. The hollow cathode device. (**a**) Illustration of the hollow cathode device to ignite the nitrogen plasmas in the inside of nozzles. (**b**) High densification in the hollow cathode.

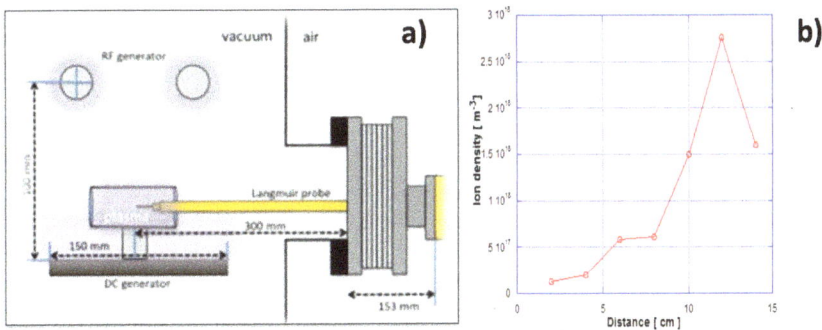

Figure 3. Langmuir probe measurement in the inside of hollow. (**a**) Experimental setup for measurement. (**b**) Ion density distribution from the inlet to the outlet.

2.3. Specimen

Besides for the normal AISI316 (Tokai Engineering Service Co. Ltd., Tokyo, Japan) pipe unit with the inner diameter of 16.9 mm, both the AISI316 mini- and AISI304 micro-nozzles were prepared to make plasma nitriding with use of the hollow cathode device. Table 1 lists their dimensional sizes and geometries, respectively. The finished inner surfaces by mechanical machining and end-milling was nitrided for pipe and nozzle specimens only after cleaning. The AISI316 austenitic stainless steel pipe was first used to describe the hollow cathode effect on the hardening behavior along its inner surface.

Both the mini- and micro-nozzles were also utilized to selectively harden the through-hole surfaces. Scanning Electron Microscopy (SEM) was utilized to describe the nitrided layer in their inner surfaces.

Table 1. Geometry, size and dimension of pipe, mini- and micro-nozzles to be processed by the present plasma nitriding.

Pipe and Nozzle	Length	Outer Diameter at Outlet	Thickness at Outlet	Inner Diameter at Outlet	Surface Roughness Rz	Outlook	Material Supplier
AISI304 Micro-Pipe	30.0 mm	0.88 mm	0.15 mm	0.58 mm	0.5 μm		TECDIA Co., Ltd.
AISI316 Pipe	55.5 mm	21.5 mm	2.3 mm	16.9 mm	1.5 μm		Tokai Engineering Service, Co., Ltd.
AISI316 Mini-Nozzle	25.0 mm	2.8 mm	0.86 mm	1.08 mm	1.2 μm		Tokai Engineering Service, Co., Ltd.
AISI304 Micro-Nozzle	5.5 mm	1.9 mm	0.5 mm	1.0 mm	0.5 μm		TECDIA Co., Ltd.

3. Experimental Results

3.1. Plasma Nitriding of AISI316 Stainless Steel Pipe

In a manner similar to the hollow cathode device in Figure 2a, the nitrogen ion density is expected to increase along the carrier gas flow direction through the pipe and nozzle since the nitrogen atom and NH–radical flux intensity is enhanced toward the outlet of the pipe and nozzle. That is, the amount of nitrogen interstitial atoms into the depth of the pipe and nozzle thickness is expected to increase from their inlet to the outlet. First, AISI316 austenitic stainless steel pipe is employed to investigate this hollow cathode effect on the nitriding behavior. With an increase in ion density, the nitriding process is enhanced by increasing the nitrogen atoms and NH–radical flux into the depth of the pipe thickness. This enhancement results in a high hardness.

Figure 4 compares the measured hardness distributions from the inlet to the outlet of the AISI316 pipe before and after plasma nitriding at 693 K for 14.4 ks by 70 Pa. A specimen with a length of 16.9 mm was cut off from the end of the outlet in this pipe after nitriding. The average hardness of nitrided through-hole inner surfaces is 800 HV higher than the matrix hardness of 200 HV at every position.

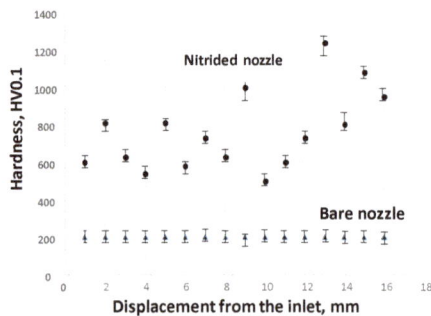

Figure 4. Comparison of the measured hardness distribution toward the end of the outlet in the normal nozzle through-hole before and after the plasma nitriding process.

In correspondence with the monotonic increase of ion density in Figure 3b, the measured hardness also increases toward the outlet of the nozzle. This implies that a through-hole of the pipe works as a hollow cathode device, and only its inner surface is selectively nitrided.

3.2. Plasma Nitriding of AISI304 Mini-Pipe

The AISI304 mini-pipe specimen was employed to make SEM and Energy Dispersive Spectroscopy (EDX) analyses on the microstructure of plasma-nitrided inner surfaces. The nitriding condition was the same as that used in the case of the AISI316 pipe. The nitrided mini-pipe at 673 K for 7.2 ks by 70 Pa was cut in half; the specimen for analysis was prepared by mechanical and chemical polishing. Figure 5 depicted the SEM image and nitrogen mapping by EDX, respectively. The surface condition did not change significantly from the initial smooth surface with a roughness around 0.5 μm. The nitrogen was uniformly distributed on the inner surface of the pipe; the nitrogen content became around 5 mass %. This means that the inner surface of the pipes should be nitrided to have a nitrogen content value that is higher than the maximum nitrogen solubility limit of 0.1 mass %.

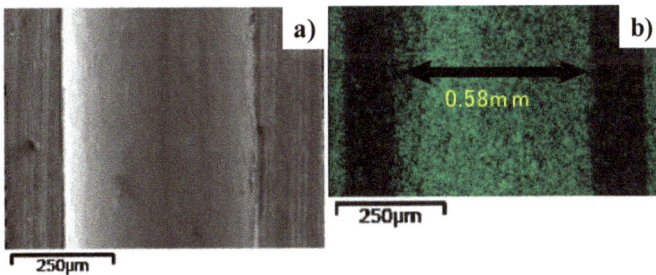

Figure 5. The plasma-nitrided AISI304 mini-pipe. (**a**) Scanning Electron Microscopy (SEM) image of the inner surface of nitrided pipe. (**b**) Nitrogen mapping on the inner surface.

3.3. Plasma Nitriding of the Mini-Nozzle

The mini-nozzle in Table 1 is widely utilized as a guide structure to control the movement of the fibers. The hardening of its inner surface is essential to prolong its lifetime. This mini-nozzle has a through-hole, the inner diameter of which decreases to 1.3 mm at the outlet. After plasma nitriding at 693 K for 14.4 ks, the nitrided mini-nozzle was wire-cut in half via electric discharge machining. The micro-Vickers testing was performed with a diamond indenter. Figure 6 depicts the optical microscope image with the hardness measurement locations from #1 to #8 along the nozzle outlet.

Figure 6. Outlet of mini-nozzle after plasma nitriding at 693 K for 14.4 ks.

As listed in Table 2, the hardness at #2 and #6 away from the through-hole inner surface is lower than 400 HV; this might be because the plasma density becomes lower than that at the vicinity of the outlet of the nozzle. In fact, the hardness values at #1 and #2 differ from each other even when both points are located closer; the inhomogeneous nitriding behavior must be attributed to a lower plasma density.

Table 2. Hardness distribution at the vicinity of the inner surface of the through-hole in the mini-nozzle outlet.

Position in Figure 4	Micro-Vickers Hardness
#1	530 HV ± 15 HV
#2	390 HV ± 20 HV
#3	660 HV ± 10 HV
#4	750 HV ± 15 HV
#5	820 HV ± 20 HV
#6	380 HV ± 10 HV
#7	690 HV ± 20 HV
#8	600 HV ± 15 HV

On the other hand, the hardness, measured at the inner surface area of the nozzle outlet, reached 600–800 HV. This proves that the inner surface of the through-hole in the mini-nozzle is nitrided and hardened at a significant depth.

Figure 7 shows the SEM image of a cross section of the mini-nozzle. Based on a noticeable difference in microstructure, the affected layer thickness by this nitriding was estimated to be 125 μm.

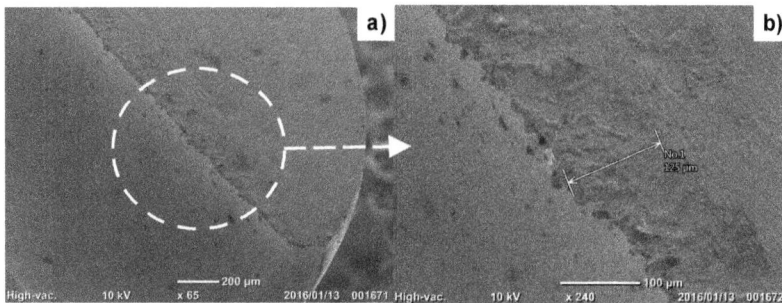

Figure 7. SEM image of the cross section for the outlet mini-nozzle thickness. (**a**) The nitrided layer is formed along the through-hole in the low magnification SEM image. (**b**) Its thickness is estimated to be 125 μm in the high magnification SEM image.

Based on Table 2, a hardness higher than 600 HV was measured at #3–5 and #7–8, in the vicinity of the top of mini-nozzle outlet. This implies that the inner surface of the through-hole of the mini-nozzle was plasma-nitrided to the extent that a hardness much higher than 200 HV for the bare AISI316 stainless steels was attained.

3.4. Plasma Nitriding of Micro-Nozzle

The micro-nozzle, which was frequently used to join electric parts, had a thin and narrow through-hole, the inner diameter of which was 200 μm at the outlet. Since the metallic melts, the solders, and the chemically active agents flow along the inside, its inner surface must be hardened and chemically modified.

In the nitriding process, the specially designed jig was prepared to fix this micro-nozzle so that the mixed gas flow direction coincides with the center line of the through-hole of the micro-nozzle. In a manner similar to the post-treatment of the mini-nozzle, the specimen for analysis was prepared from the plasma-nitrided micro-nozzle. Figure 8 depicts the optical microscopic image of the specimen with hardness measurement locations from #1 to #3 at the vicinity of the nozzle outlet inner surface. The positions of #1 and #2 are located at the vicinity of the inner surface of the through-hole, while #3 is selected to locate only on the inner surface of the through-hole.

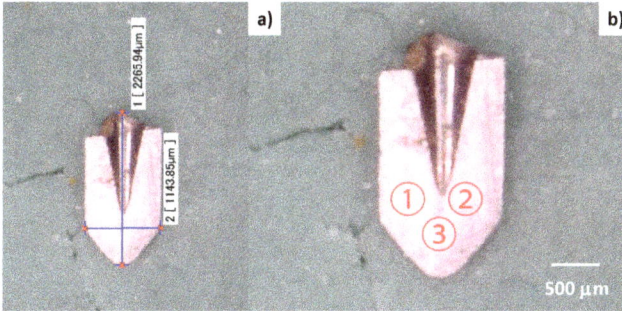

Figure 8. A cross section of the plasma nitride micro-nozzle at 693 K for 14.4 ks. (**a**) Hardness testing specimen cut-off from the original micro-nozzle. (**b**) The hardness measuring points from #1 to #3 on the flattened surface in the specimen.

Table 3 summarizes the measured micro-Vickers hardness at #1 to #3. Compared to the hardness of bare AISI316, 200 HV, the inner surface of the through-hole was significantly hardened. The average hardness for this micro-nozzle was 950 HV, much higher than that for the mini-nozzle. This is because the micro-indenter included more un-nitrided area at the outlet of a mini-nozzle to lower the measured hardness in practice. On the other hand, in the case of this micro-nozzle, the nitrided area of its thickness at the outlet mainly contributes to the measured hardness.

Table 3. Hardness distribution at the vicinity of the inner surface of the through-hole in the micro-nozzle outlet.

Position in Figure 6	Micro-Vickers Hardness
#1	960 HV ± 30 HV
#2	880 HV ± 20 HV
#3	1020 HV ± 15 HV

Figure 9 shows the SEM image of a cross section of the outlet of the through-hole. Although the fine nitriding front end was not observed in the thickness, the deeper nitrided layer might be formed by the present hollow cathode-type plasma nitriding.

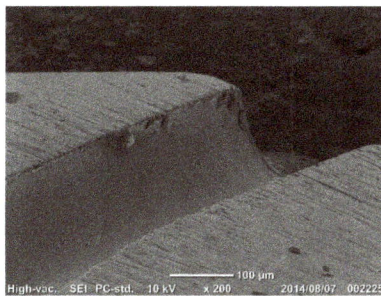

Figure 9. SEM image of a cross section at the outlet of the micro-nozzle with an inner diameter of 200 μm.

4. Discussion

Several surface modification processes are thought to improve the mechanical and chemical properties of through-hole surfaces in stainless steel pipes and nozzles. It is difficult to blow the resource gases into the through holes of pipes and nozzles for the hard coating and to stimulate the physical or chemical deposition processes. With respect to case hardening and the nitriding in gaseous and liquid phases, the entire surfaces were affected by these processes; in addition, little modification took place on the inner surfaces. In normal plasma nitriding of austenitic stainless steels, a holding temperature higher than 800 K was necessary, even for nitride precipitation hardening. That formation of precipitates has a possible risk of deteriorating the inner surface quality of through-holes in miniature nozzles.

In the present high-density, low-temperature nitriding, the inner surface of nozzles and pipes is selectively nitrided by the hollow-cathode effect. As shown in Figures 7 and 9, the nitrided layer thickness reaches to 1/8 to 1/4 of the micro-nozzle thickness. Just as reported in [8,19], the nitrided layers with thicknesses around 75–80 μm were formed by the RF/DC and RF plasma nitriding at 673 K. In particular, the highly densified plasma by the hollow-cathode effect provides a means of forming a thick nitrided layer at the inner surface of the narrow channels in the mini- and micro-nozzles. In addition, as recently pointed out in [8,11], the nitrided layer in the stainless steel matrix has a refined microstructure with an average grain size of less than 0.1 μm. This implies that the nitrided micro-nozzle has a composite structure where its inner part is nitrided to have a hardness of up to 1000 HV and high strength, and the other outer sections in the thickness remains a bare stainless steel with its intrinsic ductility and toughness. This composite structure has hardness–toughness balancing for preserving the integrity of micro-nozzles to make flushing out the viscous solders and polymer melts from its outlet to achieve joining and drawing.

Toward the industrial application of this technology, the fixture for introducing the mixture gas into the micro-nozzle needs to be first redesigned to nitride the tens or hundreds of micro-nozzles simultaneously. As keenly stated in [18], the distributed nitriding system using the hollow cathode in the cascade provides a means of performing simultaneous nitriding. Each nitriding unit in the cascaded system works as a miniature nitriding reactor with a substantial reduction of processing time in heating and cooling. Mock-up testing has recently begun as part of a METI program to nitride 36 micro-nozzles by a single process.

5. Conclusions

Low-temperature high-density plasma nitriding was proposed to make surface modifications to mini- and micro-nozzles. The inner surface of through-holes in these nozzles was sufficiently hardened, creating a hardness higher than that of the bare AISI316 by three to five times. This solution can be applied to improve the strength and hardness of manifold nozzles and to replace ceramic outlet nozzles. Furthermore, this surface modification is also capable of strengthening fine channels and miniature reservoirs in Micro Electro Mechanical Systems (MEMS).

Acknowledgments: The authors would like to express their gratitude to Shuichi Kurozumi (SIT, Tokyo, Japan), Ryo Kubota (SIT), Naoya Hayashi (SIT), and Saki Hashimoto (TECDIA, Co. Ltd., Tokyo, Japan) for their help in the experiments. This study was financially supported in part by the Ministry of Economics, Trades and Industry (METI) project, 2015.

Author Contributions: T. Aizawa and K. Wasa conceived and designed the experiments; T. Aizawa performed the experiments and wrote the paper; T. Aizawa and K. Wasa analyzed the data.

Conflicts of Interest: The authors declare no conflict of interest.

References

1. TecDia. Catalogue of Nozzles. 2016. Available online: http://www.tecdia.com/jp/products/precision/dispenser/index.php (accessed on 10 May 2017).

2. Resigen, U.; Scheik, S. Multi-dimensional line dispensing of unfilled adhesives. *Microsyst. Technol.* **2008**, *14*, 1895–1901. [CrossRef]

3. Katoh, T.; Aizawa, T.; Yamaguchi, T. Plasma assisted nitriding for micro-texturing martensitic stainless steels. *Manuf. Rev.* **2015**, *2*, 2. [CrossRef]

4. Aizawa, T.; Fukuda, T.; Morita, T. Low temperature high density plasma nitriding of stainless steel molds for stamping of oxide glasses. *Manuf. Rev.* **2016**, *3*, 2. [CrossRef]

5. Windajanti, J.M.; Aizawa, T.; Djoko, H.S. High density plasma nitriding of pure titanium. In Proceedings of the 10th SEATUC Conference 2016, Tokyo, Japan, 22–24 February 2016.

6. Aizawa, T.; Muraishi, S.; Sugita, Y. High density plasma nitriding of Al-Cu alloys for automotive parts. *J. Phys. Sci. Appl.* **2014**, *4*, 255–261.

7. Kim, S.K.; Yoo, J.S.; Priest, J.M.; Fewell, M.P. Characteristics of martensitic stainless steel nitrided in a low-pressure RF plasma. *Surf. Coat. Technol.* **2003**, *163*, 380–385. [CrossRef]

8. Farghali, A.; Aizawa, T. Phase transformation induced by high nitrogen content solid solution in the martensitic stainless steels. *Mater. Trans.* **2017**, *58*, 697–700. [CrossRef]

9. Samandi, M.; Shedden, B.A.; Smith, D.I.; Collins, G.A.; Hutchings, R.; Tendys, H. Microstructure, corrosion and tribological behavior of plasma immersion ion-implanted austenitic stainless steel. *Surf. Coat. Technol.* **1993**, *59*, 261–266. [CrossRef]

10. Blawert, C.; Mordike, B.L.; Jiraskova, Y.; Schneeweiss, O. Structure and composition of expanded austenite produced by nitrogen plasma immersion ion implantation of stainless steels X6CrNiTi1810 and X2CrNiMoN2253. *Surf. Coat. Technol.* **1999**, *116*, 189–198. [CrossRef]

11. Farghali, A.; Aizawa, T. High nitrogen concentration on the fine grained austenitic stainless steels. In Proceedings of the 11th SEATUC Conference, Ho Chi Minh City, Vietnam, 13–14 March 2017; in press.

12. Aizawa, T.; Sugita, Y. High density RF-DC plasma nitriding of steels for die and mold technologies. *Res. Rep. SIT* **2013**, *57*, 1–10.

13. Anzai, M. *Surface Treatment for High Qualification of Dies and Molds*; Nikkan-Kougyo Shinbun: Tokyo, Japan, 2009.

14. Dong, H. S-phase surface engineering of Fe-Cr, Co-Cr and Ni-Cr alloys. *Int. Mater. Rev.* **2010**, *55*, 65–98. [CrossRef]

15. Santjojo, D.J.; Istiroyah; Aizawa, T. Dynamics of nitrogen and hydrogen species in a high rate plasma nitriding of martensitic stainless steel. In Proceedings of the 9th SEATUC Conference, Nakhon Ratchasima, Thailand, 27–30 July 2015.

16. Yunata, E.E. Characterization and Application of Hollow Cathode Oxygen Plasma. Ph.D. Thesis, Shibaura Institute of Technology, Tokyo, Japan, 2016.

17. Czerwiec, T.; Michel, H.; Bergmann, E. Low-pressure, high-density plasma nitriding: Mechanisms, technology and results. *Surf. Coat. Technol.* **1998**, *108–109*, 182–190. [CrossRef]

18. Aizawa, T.; Sugita, Y. Distributed plasma nitriding systems for surface treatment of miniature functional products. In Proceedings of the 9th 4M/ICOMM, Milan, Italy, 31 March–2 April 2015; pp. 449–453.

19. Ferreira, L.M.; Brunatto, S.F.; Cardoso, R.P. Martensitic stainless steels low-temperature nitriding: Dependence of substrate composition. *Mater. Res.* **2015**, *18*, 622–627. [CrossRef]

micromachines

MDPI

Article

3D Finite Element Simulation of Micro End-Milling by Considering the Effect of Tool Run-Out

Ali Davoudinejad [1,*], Guido Tosello [1], Paolo Parenti [2] and Massimiliano Annoni [2]

[1] Department of Mechanical Engineering, Technical University of Denmark, Building 427A, Produktionstorvet, 2800 Kgs. Lyngby, Denmark; guto@mek.dtu.dk

[2] Mechanical Engineering Department, Politecnico di Milano, Via La Masa 1, 20156 Milan, Italy; paolo.parenti@polimi.it (P.P.); massimiliano.annoni@polimi.it (M.A.)

* Correspondence: alidav@mek.dtu.dk; Tel.: +45-45254897

Academic Editor: Stefan Dimov
Received: 26 April 2017; Accepted: 12 June 2017; Published: 16 June 2017

Abstract: Understanding the micro milling phenomena involved in the process is critical and difficult through physical experiments. This study presents a 3D finite element modeling (3D FEM) approach for the micro end-milling process on Al6082-T6. The proposed model employs a Lagrangian explicit finite element formulation to perform coupled thermo-mechanical transient analyses. FE simulations were performed at different cutting conditions to obtain realistic numerical predictions of chip formation, temperature distribution, and cutting forces by considering the effect of tool run-out in the model. The radial run-out is a significant issue in micro milling processes and influences the cutting stability due to chip load and force variations. The Johnson–Cook (JC) material constitutive model was applied and its constants were determined by an inverse method based on the experimental cutting forces acquired during the micro end-milling tests. The FE model prediction capability was validated by comparing the numerical model results with experimental tests. The maximum tool temperature was predicted in a different angular position of the cutter which is difficult or impossible to obtain in experiments. The predicted results of the model, involving the run-out influence, showed a good correlation with experimental chip formation and the signal shape of cutting forces.

Keywords: micro milling; finite element; run-out; chip formation; cutting force; cutting temperature; 3D simulation; measurement

1. Introduction

Micro milling is one the prevalent micro manufacturing processes in terms of a high volume and low production cost in comparison to other processes to achieve high-precision three dimensional (3D) products. Micro milling is characterized by the mechanical interaction of a sharp tool with the workpiece material, causing fractures inside the material along defined paths, eventually leading to the removal of the workpiece material in the form of chips [1]. Micro machining presents numerous challenges that have been attracting the attention of both scientific and industrial research for the past three decades [2,3]. Machinability studies are mostly based on physical experimental testing and are used as supplementary validation for the modeling approach. In micro machining, due to the process downscaling, experimental testing has to tackle several complexities such as thermal transient, tool wear, tool run-out, fixture, and workpiece, etc. [3–5]. In micro milling, some undesired phenomena such as tool wear/breakage, surface location errors, burr formation, and chatter are highly dependent on the cutting process itself. The cutter run-out also influences the process. Consequently, the prediction capability and accuracy of models is critical in simulations. Even a few microns of cutter run-out can significantly affect the accuracy of micro end-milling and create extensive force variations as opposed to conventional milling [6,7].

During the last four decades, numerical modeling (Finite Element Modeling (FEM) methods) has been recognized as a viable method for reducing or eliminating trial and error approaches in the design and optimization of machining processes. The 2D FE model investigated the effects of three microgroove parameters (groove width, edge distance, and depth ratio) on the friction and wear of textured and non-textured tools. It was revealed that microgrooves on the rake face of a cutting tool perpendicular to the chip flow direction were effective in a way that the frictional behavior at the chip–tool interface was improved. Microgrooves were also very effective in reducing flank wear and crater wear [8].

This study reported a significant effect of temperature on the flow stress that affected the cutting forces prediction results. A strain gradient plasticity based-FE model of orthogonal micro cutting of Al5083-H116 was used to investigate the influence of the tool edge radius on the size effect [9]. A micro orthogonal cutting simulation at small uncut chip thickness levels was presented to predict a sizeable strain gradient strengthening effect that was validated through micro cutting experiments [10]. Irfan et al. investigated the 2D FE modeling of Inconel 718 micro milling for different feed rates and cutting tool edge angles. A higher cutting temperature was observed at a higher feed rate and with a negative edge angle. The maximum stress distribution occurred in the contact zone between the tool and chip [11].

Afazov et al. [12] investigated micro milling force calculations using orthogonal FE methods including a mathematical model for determining the uncut chip thickness in the presence of run-out effects for AISI 4340 steel. The uncut chip thickness and the tool path trajectories were determined for distinct micro milling parameters such as the cutting velocity, cutting tool radius, feed rate, and number of teeth. The prediction of 2D FEM orthogonal micro cutting forces by considering the chip load and tool edge radius effect was utilized for the micro end milling of brass 260 in [13]. Tool trajectories, the edge radius, and run-out effects are considered in the prediction of milling forces. The cutting force coefficients were identified from a set of simulations at a range of cutting edge radii and chip loads. Another study investigated coated and uncoated cubic boron nitride (CBN) tools with the 2D finite element method (FEM) to predict the chip formation, cutting forces, temperatures, and wear rates generated in the micro milling of Ti-6Al-4V titanium alloy. The effect of run-out was not considered in the model [14]. 2D FEM micro-end milling of Ti-6Al-4V titanium alloy was investigated with a plane strain-based orthogonal cutting force model with a tool edge radius effect to validate the cutting force results [15].

Although 2D FE simulations offer some distinct advantages when studying the orthogonal cutting process, 3D FE models provide more realistic oblique configurations, mainly in milling processes with complex cutting tool geometries [16,17]. 3D FE simulations present supplementary analysis capabilities to investigate the effect of the helix angle and tool edge radius on chip flow and burr formation, which are almost impossible to be considered by 2D FE models [18].

A 3D FE model was developed to investigate the correlation between the cutting temperature and cutting edge radii in the micro-end milling of Al2024-T6 [19]. In this study, the simulation results revealed that increasing the tool edge radius increases the cutting temperature, reaching the highest temperature of 57.5 °C at the edge radius. Another study investigated 3D FEM to predict tool and workpiece temperature fields in the micro milling process of Ti-6Al-4V under various cutting conditions. A thermocouple was used for the experimental temperature measurements to validate the simulation results [20]. A 3D FEM study was developed on the same Ti-6Al-4V material in order to investigate the cutting forces in three directions under different cutting conditions, and the chip evolution and morphologies of different cutting parameters were also analyzed. The predicted and experimental chip morphologies were compared and a good agreement was observed. In terms of cutting forces, the predicted cutting forces showed a good correlation with the experimentally measured data with a 6–8% prediction error in the X- and Y-directions and a 12–15% prediction error in the Z-direction [21].

A study was conducted using 3D FEM simulations to evaluate the cutting force predictions for different cutting conditions in the micro-end milling of Al6061-T6 and the results were compared against the experiments presented in [22]. 3D FE simulations were utilized to predict the chip flow and shape during micro end-milling of Ti-6Al-4V titanium alloy in [23]. The model was developed for studying the tool wear along the micro tool and investigating the influence of the cutting edge roundness increment on the machining process performance.

Even though many studies on micro end milling have demonstrated the capability to provide information on chip formation, cutting forces, tool stresses, and temperature distribution, very limited work in the scientific literature deals with FE simulation by considering the effect of tool run-out in an integrated FE model. In the aforementioned studies, run-out was considered by combining mathematical models or other methods to the orthogonal FE model in the case of complex micro end milling operations [12,13]. However, in this study, run-out was considered in an integrated model to carry out the simulation.

In order to better understand the cutting mechanism and the effects of run-out in micro milling, 3D simulations are required, but a very limited number of studies in the literature deal with micro milling 3D simulations. However, at a macro scale, a few examples are available [16,17,21,24] that can be used as a starting point to develop a 3D micro milling model. In the literature, there is still a significant gap in our understanding of the effect of run-out in micro milling with 3D integrated finite element modeling techniques validated with accurate experimental campaigns.

The research presented in the current paper includes a 3D FE model of the micro end-milling process applied to different cutting parameters considering the effect of tool run-out and the tool edge radius for predicting the chip formation temperature distribution and cutting forces. The FE model flow chart is presented in Figure 1.

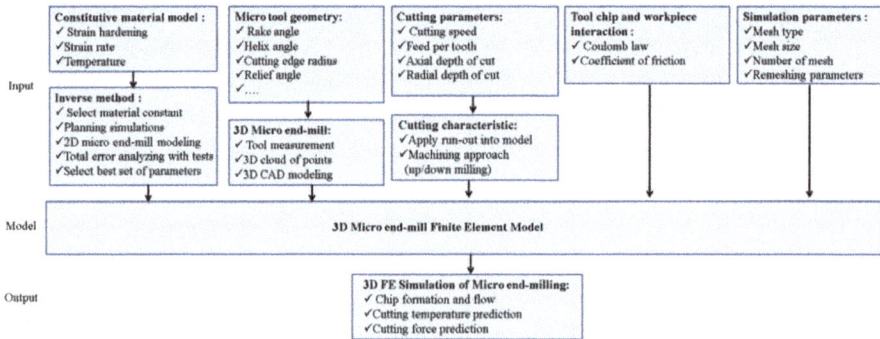

Figure 1. Flowchart for 3D micro end-milling temperature and force prediction.

The paper is structured as follows. Section 2 introduces the FE modeling and methodology, where experimental cutting forces are used to define the material model constants through an inverse method. The FE model predictions are compared against a series of micro milling experiments and the procedures are detailed in Section 3. Finally, the chip formation temperature distribution and cutting forces are discussed in Section 4. Conclusions close the paper in Section 5.

2. 3D Finite Element Simulation Setup

When dealing with FE modeling, an adequate software selection is fundamental for obtaining reliable results [25]. The 3D model for micro end-milling Al6082-T6 alloy was simulated using an explicit time integration method by employing a Lagrangian finite element formulation to perform coupled thermo-mechanical transient analysis. The AdvantEdge® FEM software (Version 6.2.011, Third Wave Systems, Minneapolis, MN, USA) was used to implement the FE model.

2.1. Tool Geometry and Modeling

The influence of the tool geometry on the chip formation, cutting temperature, tool wear, etc., was highlighted in the literature [24,26]. In this study, the selected cutting tool was a Dormer S150.05 micro flat end-mill with TiAlN-X (Titanium Aluminium Nitride Extreme) (Dormer, Sheffield, England) surface treatment. Table 1 shows the details of the nominal and actual characteristics of the tool geometry. The tools were inspected and measured by a 3D optical measuring system (Alicona Infinite Focus©) prior to machining (measurement parameters: 10× magnification, exposure time = 1.206 ms, contrast = 1, coaxial light, estimated vertical and lateral resolutions = 0.083 μm and 4 μm, respectively). The complex geometry of micro end-mills requires the use of a computer-aided design (CAD) model to obtain a complete and detailed representation of the cutter. The cutter's initial cloud of points was acquired by the Alicona Infinite Focus© system and used to precisely describe the tool geometry, taking advantage of the high density of the acquired three dimensional clouds of points. Cleaning and filtering post-processing operations after the scanning acquisition were applied to generate a reconstructed CAD model suitable for the FEM. Figure 2 shows the CAD model and cloud of points of the simulated micro end-mill.

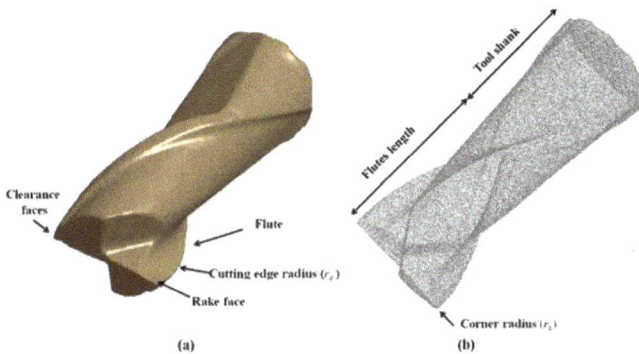

Figure 2. (**a**) Solid CAD model, (**b**) preliminary cloud of points.

Details of the actual geometry of the cutter such as the corner edge radius, helix angle, etc. (Table 1), were measured and the values extracted to reconstruct the model. The 3D STereoLithography (STL) file format was used for the FEM analysis. The tool geometry is critical to obtain reliable responses from the model. Based on the preliminary analyses [27], the nominal tool geometry was not able to reproduce the experimental force measurements.

Table 1. Nominal and actual tool characteristics. (*) the interval refers to the standard deviation/the uncertainty of the measurement.

Title	Nominal Dimensions	Actual Dimensions
Tool manufacturer	Dormer	
Code	S150.05	
Tool material	Carbide	
Surface treatment	TiAlN-X	
Flute number	2	
Diameter	500 μm	492.0 ± 2 μm (* 1.47)
Cutting edge radius (r_e)	-	3.0 ± 1 μm (* 0.71)
Helix angle	30°	27.26°
Rake angle	0°	0°
Relief angle	8°	7°
Corner radius (r_ε)	20 μm	22 μm

2.2. Cutting Configuration Setup

Figure 3 shows the setup, boundary conditions, and general geometry of the 3D FEM simulation of full slot micro end-milling. The workpiece boundary nodes were fixed in the *XY* and *Z* bottom directions and the tool constrained in the *Z* top direction, as shown in red in Figure 3a; the feed was applied by moving the tool along the *X* direction. The tool and workpiece were kept at ideal dimensions to maintain steady state cutting conditions and a minimum simulation time. The cutting tool was considered as a rigid body and the workpiece was considered as a viscoplastic material. The tool and workpiece were meshed with four node tetrahedral elements, for a total number of 56,648 and 60,916 elements, and 14,785 and 11,500 nodes, respectively, for full slot milling.

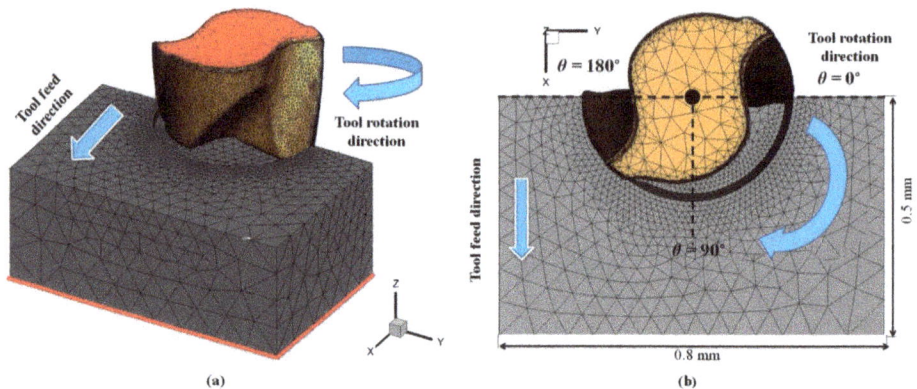

Figure 3. (**a**) 3D FEM setup perspective view, boundary condition (red area), (**b**) definition of tool engagement angle θ (top view).

After preliminary evaluation tests [28,29], the initial meshing parameters of the workpiece were set as 2 mm and 0.001 mm in terms of the maximum and minimum element size, respectively. The maximum element size of the cutter was set at 1 mm and the minimum at 0.001 mm. Other meshing parameters such as mesh grading = 0.5 (to determine the nature of transition), curvature safety = 3 (to determine the mesh accuracy), and segments per edge = 1 (to determine the density of nodes on the unit length of any edge) were selected. The choice of other meshing parameters was coherent with the recent study on meshing strategies in an FEM simulation [30]. The proper selection of meshing parameters is of great importance since it affects the accuracy of the simulation results and the computation time. A higher mesh density was considered in the area near the cutting zone in order to increase the accuracy of the computed outputs.

Adaptive remeshing was applied in order to avoid the inaccuracies due to elemental distortion, inherent to the Lagragian formulation. The mesh quality was constantly monitored during the simulations and when the element distortion reached a certain tolerance, adaptive remeshing was triggered. In addition, refinement and coarsening operators were applied in various parts of the mesh. The mesh was refined where the plastic deformation was active and coarsened in inactive regions [26].

The simulations were performed on a computer equipped with a processor characterized by 2.6 GHz, 16 cores, and 64 GB RAM. An eight threads parallel simulation mode was used to speed up the calculation time about six times. The 3D FE simulation time for a full rotation (360°) was about 25 h.

2.3. Radial Run-Out

Due to the significant effect of run-out in micro end-milling [6,7], the radial run-out (i.e., eccentricity) of the cutting tool was considered in the FE model. In the first approximation, the essential source of run-out is the cutting tool center offset [12]. The run-out definition from the ASME B89.3.4-2010 standard ("Axes of Rotation: Methods for Specifying and Testing") is "the total displacement measured by an indicator sensing against a moving surface or moved with respect to a fixed surface" [31]. This definition, applied to the case of tool run-out in milling, introduces the run-out as composed by the "spindle error motion", the tool "roundness error" (difference among cutting edge radii), and the tool-spindle "centering errors". The spindle error motion is defined as the incorrect motion of the spindle rotating axis around a reference axis; sources of error motion are bearings inaccuracies and/or the system dynamics [31].

Figure 4 represents a two-flute micro mill affected by run-out involving a centering error, indicated by the arrow between the spindle and the mill axes, and a roundness error. "Cutting edge 1", the most engaged cutting edge describing the big dashed blue circle (D_1), is affected by the roundness error that makes its radius shorter than the nominal value. The centering error and roundness error partially compensate for this in Figure 4. During the mill rotation around the spindle axis, the less engaged cutting edge, named "cutting edge 2", describes the small dashed red circle (D_2). Tool radial run-out makes the cutting edge trajectories different from the ideal identical cycloids (Figure 4); this fact makes the chip thickness produced by one cutting edge different from the other one. Consequently, the effective engagement angles (ϕ_1 and ϕ_2), the effective removed areas (A_1 and A_2), the effective widths of cut ($a_{e,1}$ and $a_{e,2}$), and the feeds per tooth ($f_{z,1}$ and $f_{z,2}$) of the two cutting edges are different in the engagements, depending on the mill centering and roundness errors.

In the developed 3D FE model, the radial run-out was considered and the center of the tool was shifted by 3 μm, corresponding to the experimental run-out measurement, along the X and Y directions of the Cartesian coordinate system. Such deviations were measured by the visual tool setter (VTS©) (Marposs, Bologna, Italy) installed on the machine (see Section 3 for details of the experimental setup).

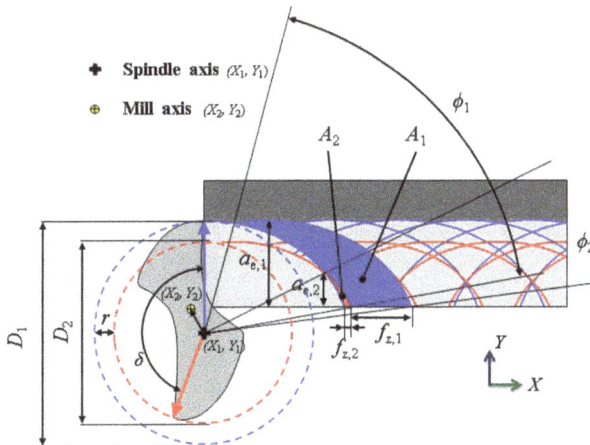

Figure 4. Definitions and cutting edge trajectories for a radial run-out affected micro end-mill.

2.4. Contact Friction Modeling

The friction phenomenon at the chip-tool interface was modeled using the Coulomb friction, as shown in Equation (1).

$$\tau = \mu\sigma_n \tag{1}$$

The frictional stresses τ on the tool rake face are assumed to be proportional to the normal stresses σ_n with a coefficient of friction μ. According to Özel [32], the sliding friction can be dominant during low cutting speed machining and the sticking friction is dominant during high-speed machining.

In micro milling, even if high rotational speeds are used, the cutting speed is lower (in this study about 31 m/min) than in macro milling due to the small tool diameter. Consequently, the Coulomb sliding friction model can be considered as effective for the micro milling process. The constant value of the friction coefficient was $\mu = 0.7$ in this study. This value was selected based on the experimental identification by Medaska on Al6061-T6 with carbide tools [33]. Al6082-T6 and Al6061-T6 are two popular aluminum alloys and sometimes they replace each other in the industrial practice due to similar characteristics.

2.5. Constitutive Material Model

The reliability of the finite element model results are greatly influenced by the material constitutive law and contact conditions at tool-chip and tool-workpiece interfaces [34]. A reliable material model that perfectly captures the constitutive behavior of the alloy under high strain, strain rate, and temperature is critical in machining simulations. A number of different constitutive material models have been reported in the literature such as Oxley's constitutive model [35], the Power law model [26], the strain path dependent model [36], and the Johnson-Cook (JC) model [37], etc. Among the different constitutive material models, JC is widely used for machining simulations due to the corresponding material behavior as a function of the strain, strain rate, and temperature [38]. In the present study, the micro end-mill is modelled as a rigid body for the coated carbide tool (TiAlN-X coating with 2 μm thickness) with the Advantedge tool default material. However, the workpiece is considered as viscoplastic material and an essential input is the accurate definition of the workpiece material properties and the constitutive material model is represented by the Johnson-Cook model. The flow stress can be expressed as (Equation (2)):

$$\sigma = \left(A + B\,(\varepsilon)^n \right)\left[1 + C\ln\left(\frac{\varepsilon'}{\varepsilon'_0}\right) \right]\left[1 - (\frac{T - T_a}{T_m - T_a})^m \right] \tag{2}$$

where σ is the material flow stress, ε is the plastic strain, ε' is the strain rate, and ε'_0 is the reference strain rate. T is the material temperature, T_m is the melting point, and T_a is the room temperature. The JC constants are as follows: A is the yield stress, B is the pre-exponential factor, C is the strain rate factor, n is the work hardening exponent, and m is the thermal softening exponent.

Inverse JC Parameters Estimation with 2D FE Simulations

The correct calibration of these parameters is critical to predict the forces, temperature, chip morphology, etc., with a reasonable accuracy. Unfortunately, there is a lack of standardization for these coefficients in the literature. The discrepancies are due to the different methods and test conditions used for the determination of the material constants [39]. To reduce the uncertainty associated with the parameters value determination, a viable method based on a machining test was adopted in this paper. It consists of an inverse method, as used by a number of researchers to obtain the material constants [39–41]. In particular, an experimental procedure for identifying the material constants at different cutting conditions was developed. An experimental campaign was specifically carried out. Table 2 shows its cutting conditions and the maximum experimental cutting force measured values. The force measuring system and compensation method applied for this set of experiments and for the final validation tests are detailed in the experimental procedure (Section 3). Each experiment was replicated two times to consider the experimental variability. Four parameters of the Johnson-Cook equation (A, B, n, C), which were the most susceptible to the strain hardening effect, were selected.

The simulations plan was based on a 2^4 full factorial design (two levels and 4 factors, Table 3). 16 JC parameter combinations were used for simulating each cutting condition of Table 2, for a total of 64 runs. An initial JC parameters combination for the 6082-T6 aluminum alloy was taken from (Table 3). These values were obtained at high strain rates by conducting split Hopkinson pressure bar (SHPB) tests in conditions similar to metal cutting. The other level was obtained by scaling down the [42] values by 50%, apart from the m one (Table 3). This scale was applied due to the overestimated result of the cutting force simulations with original values [27]. Other physical properties of Al6082-T6 used in the FE model are presented in Table 4 [43].

Table 2. Experimental and 2D simulated cutting conditions.

Cutting Conditions	Tool Diameter	Cutting Speed	Radial Depth of Cut (a_e)	Axial Depth of Cut (a_p)	Feed Per Tooth	F_x	F_y
	(mm)	(m/min)	(mm)	(mm)	(μm/(Tooth·Rev))	(N)	(N)
1	0.5	28	0.125	0.05	4	0.7	0.3
2	0.5	28	0.25	0.05	4	0.93	0.8
3	0.5	28	0.125	0.1	4	1.3	0.64
4	0.5	28	0.25	0.1	4	1.63	1.2

Table 3. Johnson-Cook constants for Al6082-T6.

Level	A (MPa)	B (MPa)	C	m	n	
1	428.5	327.7	0.00747	1.31	1.008	[39]
2	214.25	163.85	0.003735	1.31	0.504	

The FEM simulations in the inverse method were carried out to evaluate the best combination of JC parameters for obtaining the most accurate cutting forces. A simplified 2D FE model was used to reduce the computational effort to carry out the total 64 simulations, under the hypothesis that the best JC parameters would also perform well in the 3D case.

Figure 5 shows the setup and the general geometry of the 2D micro end-milling FEM simulation. The workpiece bottom boundary nodes are fixed in the Y direction and the tool in both the X and Y directions. The workpiece moves with a cutting speed or surface velocity, while the tool is stationary. The axial depth of cut is measured in the Z direction, perpendicular to the feed and speed directions. The rigid cutting tool was meshed using 670 brick elements. The workpiece was meshed with six-node quadratic triangular elements for a total number of 1566 nodes. The maximum and minimum element sizes for the workpiece were set to 0.1 mm and 0.002 mm, respectively. The maximum element size of the cutter was set at 0.2 mm and the minimum at 0.002 mm. The friction phenomenon at the chip-tool interface was the same as in the 3D FE model, with the same coefficient of friction. A large value of the interface heat transfer coefficient $h_{int} = 10^4$ (N/s mm °C) was selected for these simulations to ensure that they reached a steady state condition.

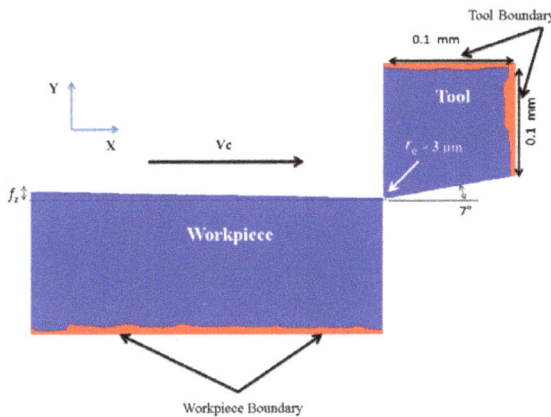

Figure 5. Setup and boundary conditions of the 2D micro end-milling FEM.

After running the 2D simulations for all the designed parameter combinations, the maximum forces were extracted from the simulations. The errors between the predicted and the experimental force components (F_x and F_y) were calculated and recorded for all tested combinations. The total error was calculated by (Equation (3)). Each relative error was squared in this equation. Squares were found and summed for each force component and each cutting condition. Subsequently, the sum of the square was divided by $M \times N$ (*M*: number of cutting conditions (4); *N*: number of force components (2)). The square root of this value gives the forces total error, representing the performance of each flow stress model representation. Table 5 shows the total errors for the 16 JC parameter combinations. In the case of the second parameters combination, an error occurred in the simulations preventing us from obtaining results in all cutting conditions.

For these sets of parameters, it was observed that the resultant total error varied between 15.73 and 73.06%. According to these results, the best set of parameters was selected to represent the material flow stress in the following 3D simulations, as presented in Table 6.

$$\text{Error}_{total} = \sqrt{\frac{\sum_{j=1}^{M} \sum_{i=1}^{N} [(\frac{F_{ji,Ex} - F_{ji,Sim}}{F_{ji,Ex}})^2]}{N \times M}} \qquad (3)$$

Table 4. Material properties for Al6082-T6 [43].

Property	Unit	Value
Young's modulus, E	(GPa)	70
Poisson ratio, v	-	0.33
Density, ρ	(g/cm^2)	2.70
Thermal conductivity, K	(W/m·K)	180
Specific heat, Cp	(J/Kg·°C)	700
Thermal expansion coefficient	-	24×10^{-6}
Melting temperature, T_{melt}	(°C)	582

Table 5. Simulated force results and comparison with experiments.

No	Coded Variable				Cutting Conditions								
	A	*B*	*n*	*C*	1		2		3		4		
					F_x	F_y	F_x	F_y	F_x	F_y	F_x	F_y	Total Error
	(MPa)	(MPa)			(N)	(N)	(N)	(N)	(N)	(N)	(N)	(N)	(%)
1	428.5	327.7	1.008	0.003735	1.4	0.33	1.77	0.4	2.4	0.7	3.2	0.82	65.56
2	214.25	163.85	0.504	0.00747	0.06	0.04	0.06	0.04	0.1	0.14	0.09	0.13	Error
3	214.25	327.7	1.008	0.003735	0.3	0.15	1.4	0.3	0.6	0.3	2.7	0.5	57.81
4	214.25	327.7	1.008	0.00747	1	0.24	1.2	0.58	0.83	0.42	2.1	0.75	31.13
5	428.5	327.7	1.008	0.00747	1.3	0.32	1.7	0.4	2.6	0.7	3.8	0.8	73.06
6	428.5	327.7	0.504	0.003735	0.82	0.43	1.1	0.57	1.7	0.88	2.4	1.1	24.7
7	428.5	163.85	0.504	0.00747	0.54	0.35	0.72	0.65	1.1	0.6	0.83	0.61	30.88
8	428.5	163.85	0.504	0.003735	0.55	0.24	0.86	0.38	0.98	0.65	1.5	0.8	28.38
9	428.5	163.85	1.008	0.00747	0.97	0.33	1.33	0.4	1.95	0.7	2.6	0.8	39.71
10	214.25	163.85	1.008	0.003735	0.65	0.16	0.08	0.14	1.4	0.3	0.45	0.2	64.24
11	214.25	163.85	1.008	0.00747	0.82	0.21	0.17	0.09	1.5	0.35	0.35	0.18	62.57
12	214.25	327.7	0.504	0.00747	0.83	0.3	0.98	0.56	1.48	0.63	1.78	0.89	15.73
13	214.25	163.85	0.504	0.003735	0.54	0.24	0.7	0.3	1.03	0.46	1.4	0.6	35.93
14	428.5	327.7	0.504	0.00747	1	0.47	1.4	0.6	2	1	2.75	1.2	47.68
15	428.5	163.85	1.008	0.003735	0.3	0.19	1.23	0.34	1.9	0.65	2.5	0.73	45.51
16	214.25	327.7	0.504	0.003735	0.85	0.27	0.65	0.54	1.6	0.53	0.92	0.83	27.87

Error: with the following material combination constants simulations did not finished to the end.

Table 6. Best set of JC Al6082-T6 flow stress model constants used for 3D simulations.

A (MPa)	B (MPa)	C	m	n	T_m (°C)	T_a (°C)
214.25	327.7	0.00747	1.31	0.504	582	21

3. FEM Validation Experimental Procedure

In order to verify the accuracy of the developed numerical model, both the predicted cutting forces and temperatures were validated against experiments. The machine tool is one of the critical aspects of micro machining in order to achieve a high precision on machined components and have a stable machining process [44]. Micro milling operations have been performed on a Kern EVO ultra precision 5-axis machining center (nominal positioning tolerance = ±1 μm, precision on the workpiece = ±2 μm). The machine tool and experimental setup are illustrated in Figure 6a.

The cutting tool tips were inspected and measured prior to machining and the only tools with minimal geometrical inaccuracies and defects were selected for the experiments. The run-out of micro end-mills was measured with both static (Figure 6c) and dynamic (Figure 6b) methods. In the static run-out measurements, a dial gauge was used by placing its probe against the tool shank, manually rotating the spindle, and reading the value of run-out as the "total indicator reading" [31]. The measured run-out was 3 ± 1 μm for both tool measurements. The micro mill dynamic run-out was measured by a visual tool setter VTS (Marposs, Bologna, Italy), is equipped with a camera synchronized with the spindle rotation to acquire a set of tool diascopic images at a defined angular step. A dedicated software analyzes these images and calculates the tool diameter, the maximum tool dynamic diameter (or flying circle diameter) D_1, and the radial runout (or TIR) r [31], corresponding to the difference between the maximum and the minimum tool dynamic radii. The minimum tool dynamic diameter D_2 can be obtained from D_1 and r (Figure 4). The experimental setup is shown in Figure 6b. The dynamic measurement was performed at the same rotational speed of the final experiments after the required VTS speed calibration. The tool radial run-out with respect to the spindle axis was 4 ± 0.5 μm and this value was used in the simulations. The measurements were repeated five times in order to guarantee statistical consistency. The standard deviation of the measurement was XYZ μm, equivalent to $XY\%$ of the measured run-out. Micro end-mills were held in HSK32 Kern/Schaublin tool holders and a precision Schaublin collet (type D14, 74-14000) was used to reduce the clamping error and the tool run-out.

Full slot micro end-milling was carried out in dry cutting conditions. The cutting parameters are summarized in Table 7. The cutting conditions were selected according to the literature [45,46], trying to avoid significant rubbing and burr formation due to the feeds per tooth below the cutting edge radius. Four replicates were performed for each cutting condition to consider experimental variability.

Table 7. Experimental tests conditions.

Cutting Parameters	Symbol	Test1	Test2
Axial depth of cut (mm)	a_p	0.05	0.05
Radial depth of cut (mm)	a_e	0.5	0.5
Feed per tooth (μm/(tooth·rev))	f_z	8	4
Cutting speed (m/min)	v_c	31.4	31.4
Spindle speed (rpm)	n	20,000	20,000
Approach		Full slot	

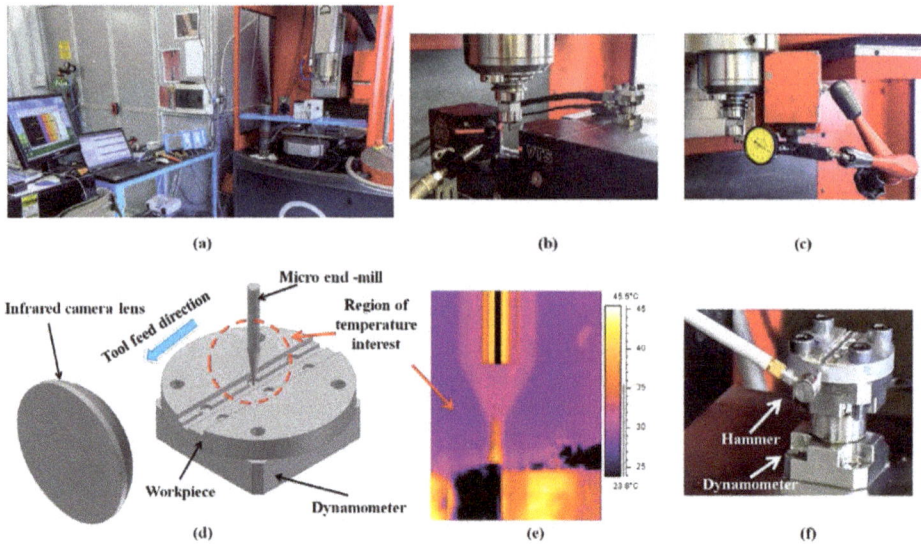

Figure 6. (a) Machine tool and experimental setup; (b) dynamic run-out measurement; (c) static run-out measurement; (d) schematic of experimental temperature measurement setup; (e) temperature distribution in end-mill; (f) dynamometer and force compensation.

3.1. Force Measurement

The cylindrical workpiece (Figure 6f) was fixed on a Kistler 9317B miniature piezoelectric three-axial dynamometer (Figure 6f), which measured the obtained cutting force signals (measuring range: F_x, $F_y = \pm 1000$ N, $F_z = \pm 2000$ N; linearity error $\leq 0.5\%$ FSO, Full Scale Output) amplified by three Kistler 5015A charge amplifiers. A low-pass filter at 20,000 Hz was directly applied on the charge amplifiers in order to avoid aliasing and a Hanning window was used to reduce leakage. The acquisition time window was one second long for each milling step. The cutting force measurements were affected by vibrations due to the low resonance frequency of the fixturing and force measurement system. In order to accurately measure the micro milling forces, a compensation method was applied to the measured forces [47]. The dynamometer dynamic behavior was identified from the impact tests applied in both the X and Y directions on the same workpiece material (Figure 6f). Figure 7 shows the result of the impact test used to obtain the frequency response function of the dynamometer in both directions. The same experimental setup and procedure was used for the inverse method experiments for determining the JC parameters (Section "Inverse JC Parameters Estimation with 2D FE Simulations").

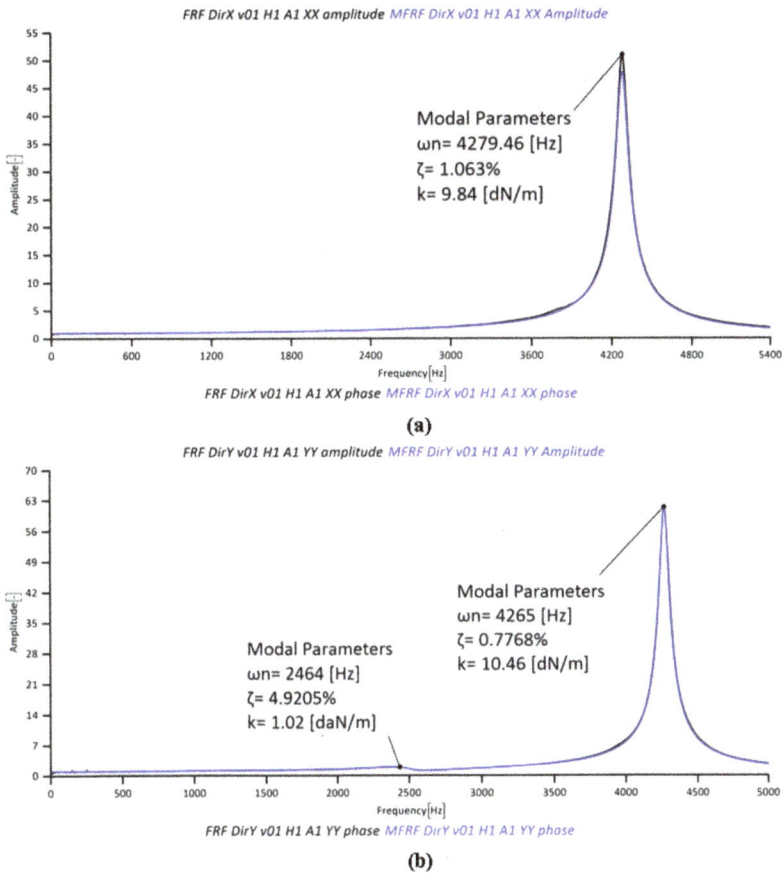

Figure 7. FRFs of the dynamometer in (**a**) X and (**b**) Y directions.

3.2. Infrared Temperature Measurements

The infrared ThermaCAM™ SC 3000 Researcher camera (FLIR Systems, Boston, MA, USA) with a waveband in the electromagnetic spectrum (in the range of 8–9 μm) was used for measuring the temperature at the cutting area. The measurement accuracy of the camera was ±1% or ±1 °C (for measurement ranges up to +150 °C) with a spatial resolution (IFOV) of 1.1 mrad. Crisp high-resolution 14-bit images and thermal data were captured and stored at high rates (up to 900 Hz) on high capacity PC. The camera was positioned along the cutting direction (Figure 6d). The temperature measurement was carried out at the tool tip area, as shown in Figure 6e. An infrared camera was calibrated with a set of standard blackbodies at various temperatures. The lenses have their own unique calibration to insure accuracy and the camera will automatically identify the lens attachment and load appropriate data in the radiometric calculations. The emissivity and background correction are variable, from 0.01 to 1.0 [48]. The camera microscope lens was used to acquire images at a 75–99 mm focus range, adjusted on the micro tool with a ±0.05 mm depth of focus. All of the experimental data have been normalized to a room temperature of 21 °C.

4. Results and Discussion

4.1. Chip Formation

The micro milling full slot operation was simulated for both types of teeth engagements. The calculation time was in the range of 30 ± 3 h, depending on the cutting condition. The different results obtained at the end of the simulations for each cutting condition are presented in this section. Figure 8 shows the chip formation and plastic strain distribution during the micro milling process in two cutting conditions, Test1 (f_z = 8 μm/(tooth·rev)) and Test2 (f_z = 4 μm/(tooth·rev)), for different angular positions of both cutting teeth. The maximum plastic strain occurred in the chip area, mainly near to the end of the cut (about 50 micros from tool tip). As expected, the plastic strain is greater when the feed is higher. The feed effect on the chip formation is also clear since different chip volumes can be observed at the same angular position. Figure 9 shows the chips at the end of the cut (θ = 360°). The effects of the tool run-out on the chip shape are visible in Figures 9 and 10. Due to the unequal load, the first and second removed chips present different shapes and sizes.

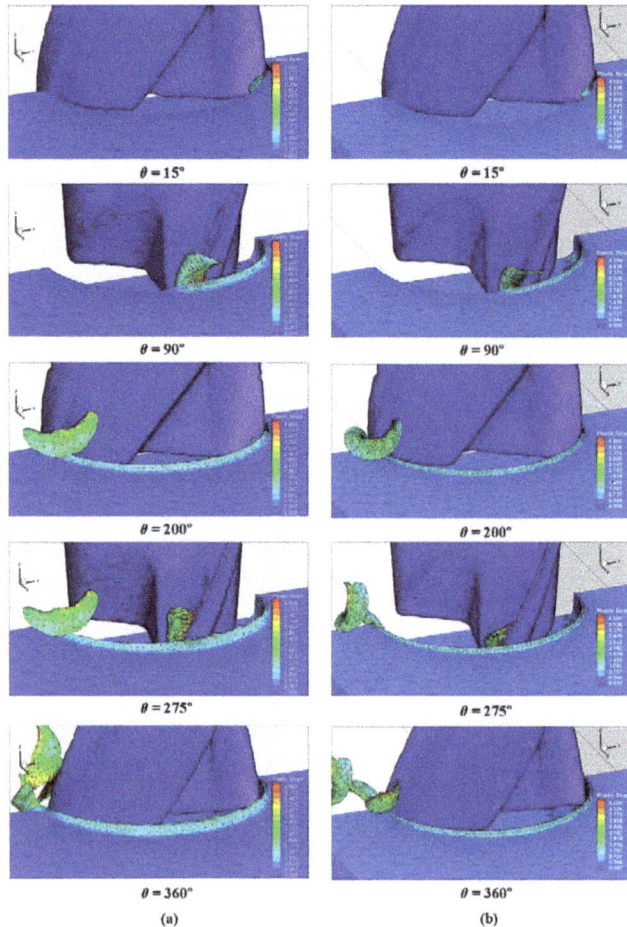

Figure 8. 3D chip formation prediction and plastic strain distribution at different angular positions (a) Test1 and (b) Test2.

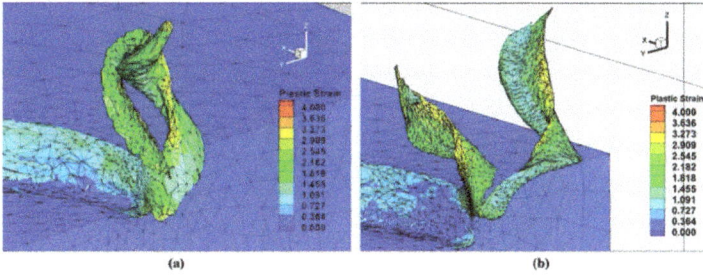

Figure 9. Run-out effect on chip formation in full slot micro milling at (**a**) Test1 and (**b**) Test2.

Figure 10 shows a comparison of the chips obtained from the experiments and simulations. According to the results of the simulations and experiments, the chips formed in the case of lower feeds (Figure 9b) tend to be curlier than in the case of higher feeds (Figure 9b). Figure 10a,c represent the biggest chips, formed by the most engaged tooth because of run-out (the first one) and Figure 10b,d represent the smallest chips, formed by the least engaged tooth (the second one). The experimental chips were collected and correspond to the simulated chip shape as they displayed different sizes and shapes.

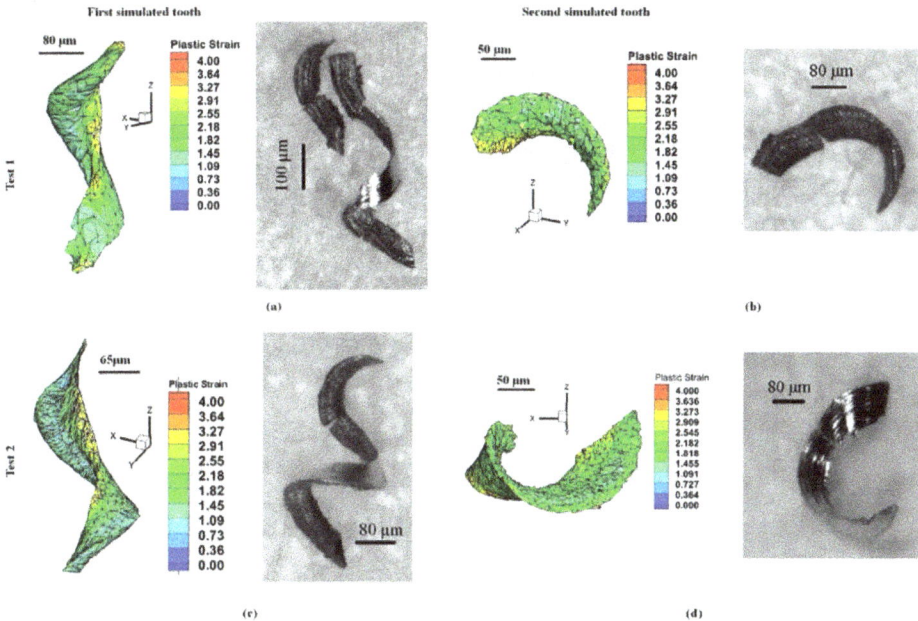

Figure 10. Comparison of predicted and acquired 3D chip shape for full slot micro end-milling of Al6082-T6, (**a**,**b**) Test1 and (**c**,**d**) Test2.

4.2. Temperature Distribution

Figure 11 shows the simulated temperature distribution in the cutting area at different angular positions of the tool with the results of two replicates of experimental cutting temperature measurements captured by the thermal camera in the full slot micro end-milling of Al6082-T6. The maximum workpiece temperatures (around 48 °C) in the simulation results can be observed at the chip area. Figure 11b shows the lower feed condition results, where a lower temperature occurs in the chip area and in the tool-workpiece contact zone. Due to the high plastic deformation taking place in the chip formation area, a higher temperature is observed in that region compared to other workpiece positions. Regarding the experimental results, the main region of interest for the temperature measurements was the tool tip, which shows the highest temperature compared to the upper part of the tool. The temperature distributions along the tool cutting edge and the tool-workpiece contact and chip area are shown in Figure 12. The variation of the highest temperature pick in different engagements toward the end of cut is presented for both cutting conditions. The tool maximum temperature region is at the corner radius and rake face of the micro cutter. The graphs of cutting tool temperature distributions (Figure 12) indicate dissimilar temperature drops after the first tooth engagement at $\theta = 200°$ and at the end of the cut caused by the feed rate variation at different cutting conditions. It can be seen that more realistic results of the temperature distributions were obtained to comprehend the effect of different cutting conditions in both the chip and cutting edge with the 3D FEM, as it is nearly impossible to monitor each cutting edge in physical experiments. Figure 13a shows the maximum temperature distribution comparison from the experimental and simulation results in the cutting area and Figure 13b presents the interaction plot from the design of experiments (DOE) analysis. The results indicate that cutting parameters influence the temperature distribution, and a lower temperature was observed in a lower cutting condition. The interaction plot shows that the simulation results were overestimated by between 7% and 12.5% in comparison with the experiments; however, a similar trend was observed with the experiments. The temperature deviation was in the range of 3–6 °C.

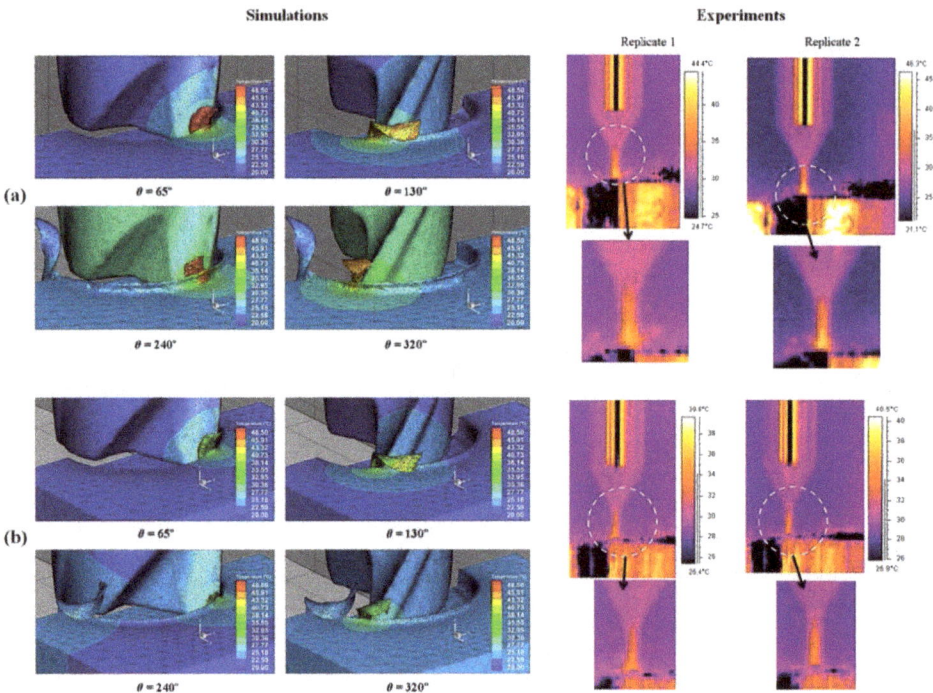

Figure 11. Simulated and experimental temperature distributions in full slot micro end-milling of Al6082-T6 at (**a**) $f_z = 8$ μm/(tooth·rev) and (**b**) $f_z = 4$ μm/(tooth·rev).

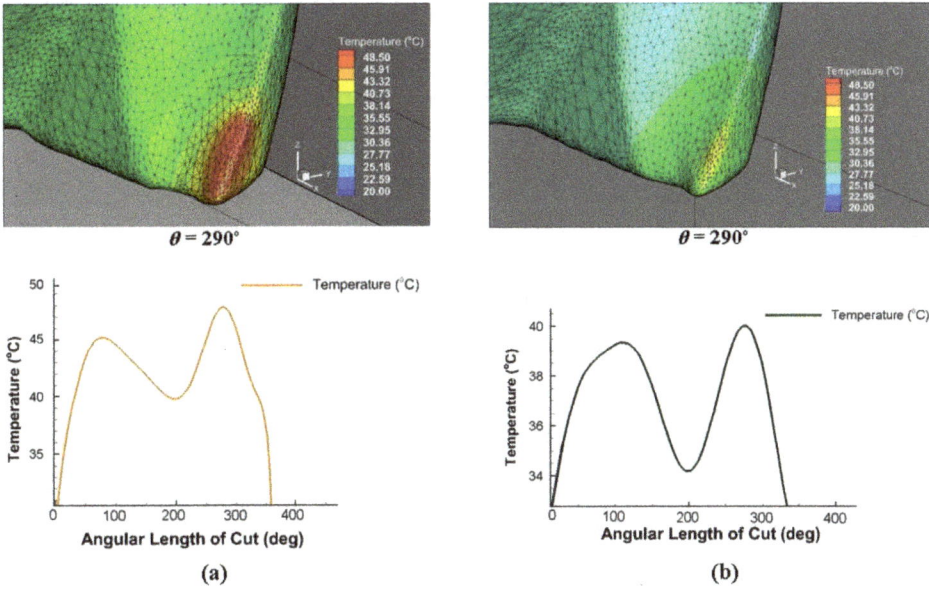

$\theta = 290°$

$\theta = 290°$

(a)

(b)

Figure 12. Simulated temperature distribution along the cutting edge for (**a**) Test1 and (**b**) Test2.

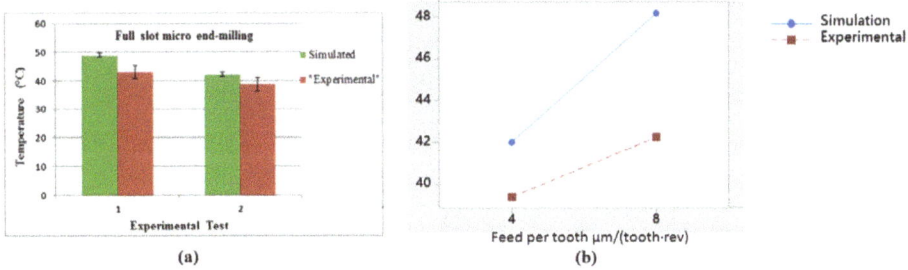

(a)

(b)

Figure 13. (**a**) Comparison of experimental and simulated maximum temperature distribution and (**b**) interaction plot of temperature distribution

4.3. Cutting Forces Predictions

It was noted that all the simulated forces were affected by a relevant amount of numerical noise distributed at an extremely high frequency range, much beyond the range of real cutting force contributions. A fifth-order Butterworth low-pass filter was therefore applied to the simulated signals with a cut-off frequency equal to the bandwidth achievable by the dynamometer installed on the machine, which is around 4000 Hz (after the application of the acquired force compensation [41]), in order to obtain simulated force signals comparable with the experiments. The comparisons were performed between results obtained by the finite element model and the physical experiments. Figure 14 shows the cutting force components F_x, F_y, and F_z obtained for both teeth engagements, considering the effect of the run-out in the full slot micro end-milling process. The comparison between the experiments and simulation reveals globally similar trends for F_x, and F_y components in terms of the curve shapes. However, some discrepancies in the magnitude of forces were observed, probably due to aluminum particles sticking to the tool (build up edge) that affect the tool geometry and the tool dynamics. The effect of the run-out phenomenon is visible in the cutting force results as the first tooth shows higher values of the F_x and F_y components due to the larger material removal. The F_z component was almost constant during the cutting operation in both the experimental and simulation results.

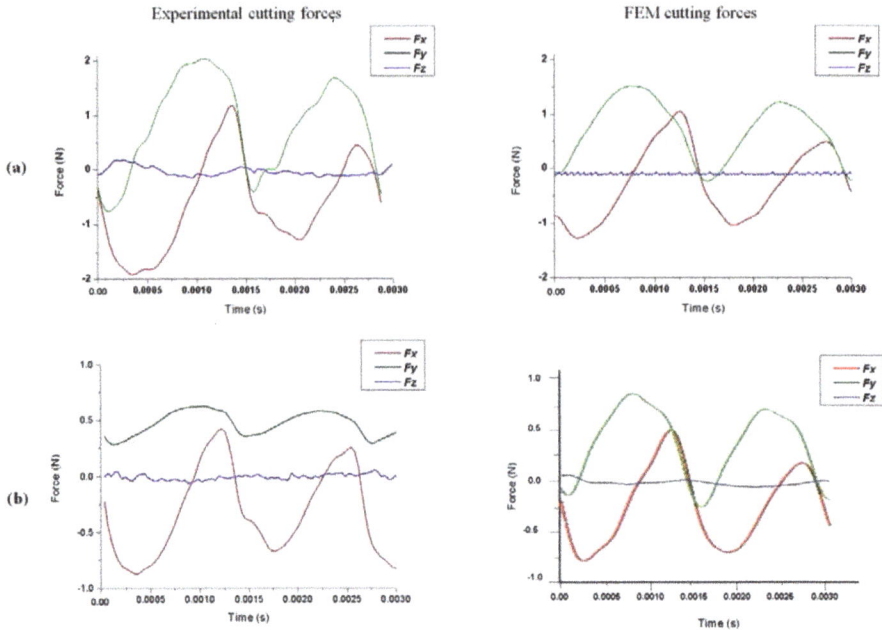

Figure 14. Experimental and FEM cutting force curves for (a) Test1 and (b) Test2.

5. Conclusions

A 3D FEM was developed in this work for studying the micro end-milling process on aluminum alloy 6082-T6. The advantages of 3D FE simulations were used with additional analyses on the effect of tool run-out. The model was applied to investigate the chip formation, cutting tool temperature distribution, and cutting forces in full slot milling. The simulation results were compared against a series of experiments. Some conclusive findings of this study are outlined in the following:

- The constants of the Al6082-T6 Johnson-Cook material constitutive model were obtained through an inverse method based on an experimental analysis on the cutting forces. Different combinations of parameters were employed to determine proper constants by comparing the experimental and simulation results.
- The chip formation was observed in realistic simulations thanks to the 3D FEM approach. An unequal chip load caused by run-out was obtained in both the experiments and simulations, together with different chip shapes for the two teeth engagements.
- Higher temperatures were observed in high feed rate conditions at the tool cutting edge radius and the maximum temperature distribution was different along the cutting edge at various angular positions of the tool. The highest temperature distribution was on the chips and tool cutting edge radius area. Experimental temperature distributions were mainly observable on the tool area and were in line with simulation results.
- The F_x, F_y, and F_z cutting force components were obtained as FEM predictions and experimental results for the two teeth, pointing out their different cutting action due to run-out.

The comparison between 3D simulations and experiments confirms that the model was capable of predicting the main features of the complex micro end-milling process on 6082-T6 aluminum alloy.

Acknowledgments: The research leading to these results has received funding from the People Programme (Marie Curie Actions) of the European Union's Seventh Framework Programme (FP7/2007-2013) under REA grant agreement No. 609405 (COFUNDPostdocDTU). The authors would also like to acknowledge the Politecnico di Milano University for founding the PhD project "3D Finite Element Modeling of Micro End-Milling by Considering Tool Run-Out, Temperature Distribution, Chip and Burr Formation" by A.D. This research work was undertaken in the context of MICROMAN project ("Process Fingerprint for Zero-defect Net-shape MICROMANufacturing", http://www.microman.mek.dtu.dk/). MICROMAN is a European Training Network supported by Horizon 2020, the EU Framework Programme for Research and Innovation (Project ID: 674801).

Author Contributions: Ali Davoudinejad carried out simulations, experiments, measurements and wrote the paper; Guido Tosello, Paolo Parenti and Massimiliano Annoni revised the paper.

Conflicts of Interest: The authors declare no conflict of interest.

References

1. Uriarte, L.; Herrero, A.; Lvanov, A.; Oosterling, H.; Staemmler, L.; Tang, P.T.; Allen, D. Comparison between microfabrication technologies for metal tooling. *Proc. Inst. Mech. Eng. Part C J. Mech. Eng. Sci.* **2006**, *220*, 1665–1676. [CrossRef]
2. DeVor, R.E.; Ehmann, K.F.; Kapoor, S.G. *Technology Assessment on Current Advanced Research in Micro-Machining and Related Areas*; The Association For Manufacturing Technology: McLean, VA, USA, 2004; p. 239.
3. Dornfeld, D.; Min, S.; Takeuchi, Y. Recent advances in mechanical micromachining. *CIRP Ann. Manuf. Technol.* **2006**, *55*, 745–768. [CrossRef]
4. Anand, R.S.; Patra, K. Modeling and simulation of mechanical micro-machining—A review. *Mach. Sci. Technol.* **2014**, *18*, 323–347. [CrossRef]
5. Koo, J.-Y.; Kim, J.-S.; Kim, P.-H. Machining Characteristics of Micro-Flow Channels in Micro-Milling Process. *Mach. Sci. Technol.* **2014**, *18*, 509–521. [CrossRef]
6. Bao, W.Y.; Tansel, I.N. Modeling micro-end-milling operations. Part II: Tool run-out. *Int. J. Mach. Tools Manuf.* **2000**, *40*, 2175–2192. [CrossRef]
7. Dhanorker, A.; Özel, T. An Experimental and Modeling Study on Meso/Micro End Milling Process. In Proceedings of the ASME 2006 International Manufacturing Science and Engineering Conference, Ypsilanti, MI, USA, 8–11 October 2006; pp. 1071–1079.
8. Ma, J.; Duong, N.H.; Lei, S. Finite element investigation of friction and wear of microgrooved cutting tool in dry machining of AISI 1045 steel. *Proc. Inst. Mech. Eng. Part J J. Eng. Tribol.* **2014**, *229*, 449–464. [CrossRef]
9. Liu, K.; Melkote, S.N. Finite element analysis of the influence of tool edge radius on size effect in orthogonal micro-cutting process. *Int. J. Mech. Sci.* **2007**, *49*, 650–660. [CrossRef]
10. Liu, K.; Melkote, S.N. Material strengthening mechanisms and their contribution to size effect in micro-cutting. *J. Manuf. Sci. Eng.* **2005**, *128*, 730–738. [CrossRef]
11. Ucun, İ.; Aslantas, K.; Bedir, F. Finite element modeling of micro-milling: Numerical simulation and experimental validation. *Mach. Sci. Technol.* **2016**, *20*, 148–172. [CrossRef]
12. Afazov, S.M.; Ratchev, S.M.; Segal, J. Modelling and simulation of micro-milling cutting forces. *J. Mater. Process. Technol.* **2010**, *210*, 2154–2162. [CrossRef]
13. Jin, X.; Altintas, Y. Prediction of micro-milling forces with finite element method. *J. Mater. Process. Technol.* **2012**, *212*, 542–552. [CrossRef]
14. Thepsonthi, T.; Özel, T. Experimental and finite element simulation based investigations on micro-milling Ti-6Al-4V titanium alloy: Effects of cBN coating on tool wear. *J. Mater. Process. Technol.* **2013**, *213*, 532–542. [CrossRef]
15. Pratap, T.; Patra, K.; Dyakonov, A.A. Modeling cutting force in micro-milling of Ti-6Al-4V titanium alloy. *Procedia Eng.* **2015**, *129*, 134–139. [CrossRef]
16. Bajpai, V.; Lee, I.; Park, H.W. Finite element modeling of three-dimensional milling process of Ti-6Al-4V. *Mater. Manuf. Process.* **2014**, *29*, 564–571. [CrossRef]
17. Wu, H.B.; Zhang, S.J. 3D FEM simulation of milling process for titanium alloy Ti6Al4V. *Int. J. Adv. Manuf. Technol.* **2014**, *71*, 1319–1326. [CrossRef]
18. Min, S.; Dornfeld, D.A.; Kim, J.; Shyu, B. Finite element modeling of burr formation in metal cutting. *Mach. Sci. Technol.* **2001**, *5*, 307–322. [CrossRef]

19. Yang, K.; Liang, Y.; Zheng, K.; Bai, Q.; Chen, W. Tool edge radius effect on cutting temperature in micro-end-milling process. *Int. J. Adv. Manuf. Technol.* **2011**, *52*, 905–912. [CrossRef]

20. Mamedov, A.; Lazoglu, I. Thermal analysis of micro milling titanium alloy Ti-6Al-4V. *J. Mater. Process. Technol.* **2016**, *229*, 659–667. [CrossRef]

21. Wang, F.; Zhao, J.; Li, A.; Zhu, N.; Zhao, J. Three-dimensional finite element modeling of high-speed end milling operations of Ti-6Al-4V. *Proc. Inst. Mech. Eng. Part B J. Eng. Manuf.* **2014**, *228*, 893–902. [CrossRef]

22. Davoudinejad, A.; Parenti, P.; Annoni, M. 3D finite element prediction of chip flow, burr formation, and cutting forces in micro end-milling of aluminum 6061-T6. *Front. Mech. Eng.* **2017**, 1–12. [CrossRef]

23. Thepsonthi, T.; Özel, T. 3-D finite element process simulation of micro-end milling Ti-6Al-4V titanium alloy: Experimental validations on chip flow and tool wear. *J. Mater. Process. Technol.* **2015**, *221*, 128–145. [CrossRef]

24. Ulutan, D.; Lazoglu, I.; Dinc, C. Three-dimensional temperature predictions in machining processes using finite difference method. *J. Mater. Process. Technol.* **2009**, *209*, 1111–1121. [CrossRef]

25. Bil, H.; Kılıç, S.E.; Tekkaya, A.E. A comparison of orthogonal cutting data from experiments with three different finite element models. *Int. J. Mach. Tools Manuf.* **2004**, *44*, 933–944. [CrossRef]

26. Man, X.; Ren, D.; Usui, S.; Johnson, C.; Marusich, T.D. Validation of finite element cutting force prediction for end milling. *Procedia CIRP* **2012**, *1*, 663–668. [CrossRef]

27. Davoudinejad, A. *3D Finite Element Modeling of Micro End-Milling by Considering Tool Run-Out, Temperature Distribution, Chip and Burr Formation*; University of Politecnico di Milano: Milan, Italy, 2016.

28. Rebaioli, L.; Annoni, M.; Davoudinejad, A.; Parenti, P. Performance of Micro End Milling Force Prediction on Aluminum 6061-T6 with 3D FE Simulation. In Proceedings of the 4M/ICOMM 2015-International Conference on Micromanufacturing, Milan, Italy, 31 March–2 April 2015.

29. Davoudinejad, A.; Chiappini, E.; Tirelli, S.; Annoni, M.; Strano, M. Finite Element Simulation and Validation of Chip Formation and Cutting Forces in Dry and Cryogenic Cutting of Ti-6Al-4V. *Procedia Manuf.* **2015**, *1*, 728–739. [CrossRef]

30. Niesłony, P.; Grzesik, W.; Chudy, R.; Habrat, W. Meshing strategies in FEM simulation of the machining process. *Arch. Civ. Mech. Eng.* **2015**, *15*, 62–70. [CrossRef]

31. American National Standards Institute (ASME). *Axes of Rotation: Methods for Specifying and Testing*; American Society of Mechanical Engineers: New York, NY, USA, 1986.

32. Özel, T. The influence of friction models on finite element simulations of machining. *Int. J. Mach. Tools Manuf.* **2006**, *46*, 518–530. [CrossRef]

33. Medaska, M.K.; Nowag, L.; Liang, S.Y. Simultaneous measurement of the thermal and tribological effects of cutting fluid. *Mach. Sci. Technol.* **1999**, *3*, 221–237. [CrossRef]

34. Umbrello, D.; M'Saoubi, R.; Outeiro, J.C. The influence of Johnson-Cook material constants on finite element simulation of machining of AISI 316L steel. *Int. J. Mach. Tools Manuf.* **2007**, *47*, 462–470. [CrossRef]

35. Oxley, P.L.B. *The Mechanics of Machining: An Analytical Approach to Assessing Machinability*; E. Horwood: Chichester, UK, 1989.

36. Katsuhiro, M.; Takahiro, S.; Eiji, U. Flow stress of low carbon steel at high temperature and strain rate. II: Flow stress under variable temperature and variable strain rate. *Bull. Jpn. Soc. Precis. Eng.* **1983**, *17*, 167–172.

37. Johnson, G.R.; Cook, W.H. A constitutive Model and Data for Metals Subjected to Large Strains, High Strain Rates and High Temperatures. In Proceedings of the 7th International Symposium on Ballistics, Hague, The Netherlands, 19–21 April 1983; pp. 541–547.

38. Dixit, U.S.; Joshi, S.N.; Davim, J.P. Incorporation of material behavior in modeling of metal forming and machining processes: A review. *Mater. Des.* **2011**, *32*, 3655–3670. [CrossRef]

39. Daoud, M.; Jomaa, W.; Chatelain, J.F.; Bouzid, A. A machining-based methodology to identify material constitutive law for finite element simulation. *Int. J. Adv. Manuf. Technol.* **2015**, *77*, 2019–2033. [CrossRef]

40. Guo, Y.B. An integral method to determine the mechanical behavior of materials in metal cutting. *J. Mater. Process. Technol.* **2003**, *142*, 72–81. [CrossRef]

41. Özel, T.; Zeren, E. A Methodology to determine work material flow stress and tool-chip interfacial friction properties by using analysis of machining. *J. Manuf. Sci. Eng.* **2005**, *128*, 119–129. [CrossRef]

42. Jaspers, S.P.F.; Dautzenberg, J. Material behaviour in conditions similar to metal cutting: Flow stress in the primary shear zone. *J. Mater. Process. Technol.* **2002**, *122*, 322–330. [CrossRef]

43. Aalco metals Ltd. *Aluminium Alloy 6082-T6~T651 Plate*; Aalco Metals Ltd.: Cobham, UK, 2013; pp. 24–25.

44. Lamikiz, A.; de Lacalle, L.N.; Celaya, A. Machine Tool Performance and Precision. In *Machine Tools for High Performance Machining*; de Lacalle, L.N., Lamikiz, A., Eds.; Springer: London, UK, 2009; pp. 219–260.

45. Gillespie, L.K.; Blotter, P.T. The Formation and Properties of Machining Burrs. *J. Eng. Ind.* **1976**, *98*, 66. [CrossRef]

46. Mathai, G.K.; Melkote, S.N.; Rosen, D.W. Effect of process parameters on burrs produced in micromilling of a thin nitinol foil. *J. Micro Nano Manuf.* **2013**, *1*, 21005–21010. [CrossRef]

47. Annoni, M.; Pusterla, N.; Rebaioli, L.; Semeraro, Q. Calibration and validation of a mechanistic micromilling force prediction model. *J. Manuf. Sci. Eng.* **2015**, *138*, 11001–11012. [CrossRef]

48. *ThermCAM® Researcher Features/Specifications*; FLIR Systems: Boston, MA, USA, 2002.

![micromachines logo] *micromachines*

MDPI

Article

Handling in the Production of Wire-Based Linked Micro Parts †

Philipp Wilhelmi *, Christian Schenck and Bernd Kuhfuss

BIME—Bremen Institute for Mechanical Engineering, MAPEX Center for Materials and Processing, University of Bremen, Badgasteiner Str. 1, 28359 Bremen, Germany; schenck@bime.de (C.S.); kuhfuss@bime.de (B.K.)
* Correspondence: wilhelmi@bime.de; Tel.: +49-421-218-64807
† This paper is an extended version of our paper presented in the 4M/IWMF 2016 Conference, Kongens Lyngby, Denmark, 13–15 September 2016, Diameter adaptive guides for wire-based linked micro parts.

Academic Editors: Hans Nørgaard Hansen and Guido Tosello
Received: 28 February 2017; Accepted: 18 May 2017; Published: 25 May 2017

Abstract: For simplified processing and the enhancement of output rate in multi-stage production, micro parts are handled as linked parts. This contribution discusses handling specific challenges in production based on an exemplary process chain. The examined linked parts consist of spherical elements linked by wire material. Hence, the diameter varies between the wire and part. Nevertheless, the linked parts must be handled accurately. The feed system is an important component too, but special focus is given to the guides in this present study. They must adapt to the diameters of both the parts and the linking wires. Two alternative variants of adaptive guides are presented and investigated under the aspects of precise radial guiding, vibration isolation, damping behavior and friction force.

Keywords: multi-stage micro part production; linked parts; diameter adaptive guides

---

1. Introduction

Handling constitutes a major challenge in the production of micro parts with regard to achieving an economic and efficient production. Due to the typically low price per unit, high outputs and consequently low cycle times are required. At the same time, precise positioning is necessary as the parts are very sensitive to mechanical damage and size effects [1]. For example, dominating adhesive forces may complicate handling. Van Brussel et al. [2] provided a broad overview of the assembly of microsystems, in which handling is a main aspect. Adhesive forces are primarily derived from electrostatic attraction, Van der Waals forces, and surface tension. These forces depend on environmental conditions and can also be used for pick-up operations, but unfortunately complicate handling. Adhesion results in inaccurate grasping and negatively affects releasing. Grasping devices in general have been addressed in a previous study [3]. Tichem et al. [4] gave a good overview of the micro-grip principles, while Fantoni et al. reviewed releasing strategies [5]. These challenges led to the development of new micro manufacturing and assembly concepts with adopted and integrated machine designs [6,7]. A complete micro bulk forming system was presented and analyzed by Arentoft et al. [8]. They considered not only the forming system itself but also all process steps, including handling. Subsequently, in another study [9] examining this bulk forming system, a new handling system that maximizes the clock rate to up to 250 strokes per minute was presented and characterized. Furthermore, Fleischer et al. [10] identified the supply of parts as a major bottleneck in micro assembly and presented a modular vibratory conveyor system, including options for separation, quality control, discard lock out and part orientation. A measure to facilitate handling and enable a production at higher output rates is the production in linked parts, which was introduced in a previous study [11]. This is especially important for micro forming processes and their special requirements [12,13], where multiple different operations may be performed on a single part. Applications of this approach to bulk forming based on sheet metal have been presented previously in the literature [14,15]. In contrast, the work presented in this paper addresses bulk forming based on wire material. Figure 1 shows an example of the production of wire-based linked micro parts. In the first process, preforms are generated within a wire by a laser material accumulation process [16]. The wire is melted partially while being pushed together. The remaining sections of the wire build the frame structure that links up the preforms. In a second step, the preforms are formed by micro rotary swaging [17,18].

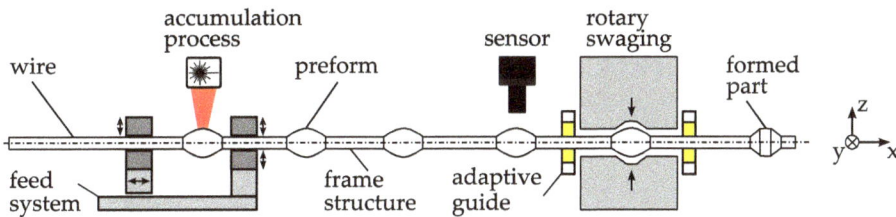

Figure 1. Example of wire-based linked micro part production.

To enable the production of linked parts at high output rates, appropriate production facilities are developed and investigated. The feed system is essential for the accumulation process. It must provide a high dynamic positioning with adequate precision and accurate alignment of the wire within the process zone. The chosen solution in this case is a gripper-based feed system driven by a linear direct drive. The grippers are self-centering and aligned to each other by the design concept. The gripper geometry provides good support of the wire by line contact. Nevertheless, the distance of the preforms within the linked parts may vary due to changes in length of the frame structure during a multi-stage production process, especially when forming is included [18]. Hence, a referencing of the individual parts becomes necessary for the positioning in a subsequent production process. Therefore, a visual high-speed part referencing system has been developed [19]. The system is based on a line camera, which is triggered based on the signal of the position sensor of the feed system. Appropriate guides are important to guide the linked parts securely through the process chain, and such guides are also addressed in this paper.

2. Requirements for Adaptive Guides

Depending on the processes and their spatial alignment, the guides along the process chain must meet different requirements. It is important that the guides suppress vibrations that occur between these processes. The part must be centered within the tools, close to the process. The vibrations induced by the process should be damped and the transmission of vibrations from the process to the linked parts outside the tool and vice versa must be suppressed. Furthermore, a compact size can be important due to space limitations. Conventional wire guides are typically realized as bushings consisting of ceramic materials. They are especially optimized for high wear resistance. As they are implemented as sliding guides, they usually feature a clearance. The clearance cannot be downscaled linearly to the micro range, which is related to tolerancing issues in micro production, as discussed in a previous study [20]. This problem already exists in the case of guiding raw wire material, because the diameter tolerances are too high. However, this becomes even more evident when considering the standard deviation of the preform diameters, which depend on the actual process parameters. Furthermore, with a constant bushing diameter, the clearance depends on the position of the parts, which can lead to a significant misalignment of the center line of the linked part (Figure 2). The linked parts used for the presented investigations feature a frame diameter of d_{frame} = 360 μm and a part diameter of d_{part} = 660 μm. Hence, assuming that there is no clearance between the guide and parts, the maximum possible misalignment is e_r = 150 μm for this example.

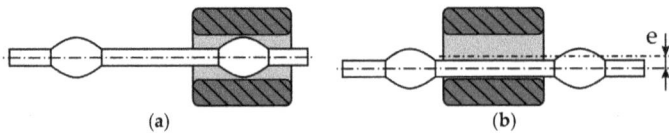

Figure 2. Misalignment in the case of applying a classical wire guide bushing: (**a**) part within the bushing; (**b**) only wire within the bushing.

The measure to overcome this problem by increasing the length of the guide beyond the maximum distance between the parts is limited by the following reasons. First, due to space limitation, this solution is not always applicable. Furthermore, the nearest guiding point in relation to the tool would depend on the position of parts in this case. Considering the low structural stiffness of the linked parts, precise guiding cannot be guaranteed this way. Adaptive guides are investigated to solve the described problems, which can locally adapt their diameter to the actual needs [21]. The required features of such a guide can be summarized as follows: precise radial guiding with sufficient stiffness, good re-centering ability after displacement of the linked parts, high damping of radial and axial vibrations, low and preferably constant resistant forces against feed direction, compactness, high wear resistance, ability of lubricant free operation, and reasonable manufacturing costs by good manufacturability. Generally, the design of micro bearings is strongly influenced by the manufacturability. Nevertheless, studies on air bearings [22,23], magnetic bearings [24,25], and ball bearings [26] can be found. All of these guiding solutions have two well-defined partners with constant geometrical properties. In the field of high accuracy requirements in combination with geometrical adaption, only gripper-based concepts are mentioned. A simpler solution with limited travel is the application of flexure hinges, which further offer the advantage of zero clearance and therefore are also applied for micromanipulation devices [27].

3. Concepts of Adaptive Guides

The general function of a linear guide is to limit the original six degrees of freedom of the guided part. In this present study, this means a translation in the feed direction and a rotation around the center line. By this definition, a wide solution space is spanned. Adaptive bearings can be either active or passive [28]. An active adaptive guide controls the clearance by measuring it with a sensor and adapting the diameter with an actuator comparable to the aforementioned grasping solutions. Regarding the needed components, such a system is comparatively complex. The complexity obviously becomes a major challenge when considering the huge number of adaptive guides that are required for the production of linked parts. Hence, a passive solution is more suitable and the presented concepts in this paper are limited to pure mechanical systems. A function scheme of a mechanical passive adaptive guide is illustrated in Figure 3. It uses the feeding force to increase the guide diameter when a part passes through. It consists of three elementary functional elements: a contact element allows relative motion and guarantees wear resistance; a kinematics transforms the feeding force to a radial force, which acts on an elastic element to enable the diameter adaption. In the macro range for high precision applications, a high stiffness is usually preferred. In the considered case, a reduced stiffness has less impact on the accuracy, due to the inherent low rigidity of the linked parts.

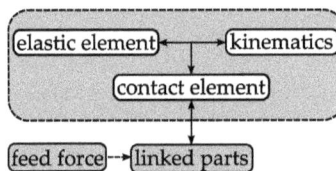

Figure 3. Passive adaptive guide function scheme.

Based on this scheme, two prototypes were built. Concept 1 (see Figure 4a) uses bearing balls and spiral springs. The kinematics is realized by the radius of the bearing ball and the contour of the parts. The advantage of this solution is that the components are available as standard parts with high quality. Based on the characteristics of the used components and the considered linked parts, the nominal preload of the guide is $F_{frame} = 0.34$ N for a frame diameter of $d_{frame} = 360$ μm. The force of one spring on a part of diameter $d_{part} = 660$ μm is $F_{part} = 0.66$ N.

The maximum simplification with respect to the number of components is realized by concept 2 (see Figure 4b). A cylindrical elastomer element with a central bore integrates all functions in one component. In order to realize the preload, the bore diameter is chosen to be slightly smaller than the frame diameter, d_{frame}. The advantage of the simplification of this concept goes along with the disadvantage of the coupling guiding features, such as stiffness and wear resistance. They depend on the applied material. In the following, the concepts are called G1 and G2.

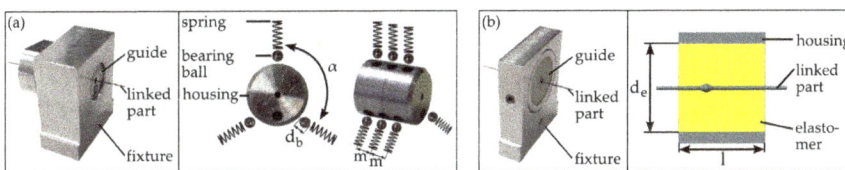

Figure 4. (**a**) Concept G1, $d_b = 2$ mm, $\alpha = 120°$, $m = 3$ mm, spring rate $R = 2.122$ mN/mm; (**b**) Concept G2, $d_e = 16$ mm, $l = 8$ mm, material: Polyurethan, 70 Shore A.

4. Measurements and Results

The requirements formulated in the previous section can be divided into two groups. Some properties of the guides, such as compactness, ability of lubricant-free operation, manufacturability, and partial wear resistance, are more or less given by the construction. Others have a direct influence on the processes and need to be determined experimentally in detail. These aspects, which are analyzed in the following, are guiding precision, influence on vibrations, and influence on the feed force.

4.1. Guiding Precision

For the evaluation of the guiding precision, the displacement in the z-direction is measured while a probe is moved through the adaptive guide (Figure 5). This is conducted in two measurement series with two different kinds of probes. First, raw wire is used, and second, the same measurements are performed with linked parts. The setup consists of a linear axis mounted on a base plate with different mounting positions for the guides.

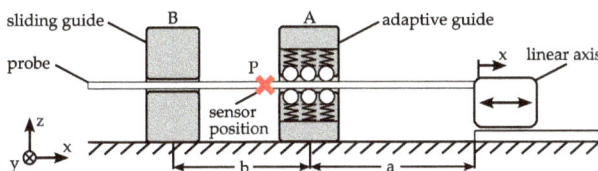

Figure 5. Measurement setup—radial deviation, $a = 142$ mm, $b = 60$ mm.

For each of the tests, one of the two adaptive guides, G1 and G2, as well as an additional sliding guide are used. Their positions are named A and B. The distances between A and the home position $x = 0$ mm of the linear axis as well as the distance between A and B are described by the variables a and b, respectively. The different adaptive guides are mounted in position A. Furthermore, an aluminium block with a bore of a diameter of $d_{sb} = 400$ μm is used for sliding the bushing at position B. The sensor is located at position P, directly outside the adaptive guide. The position of the probe in the z-direction is measured by an optical micrometer (Figure 6). The accuracy of the used device is specified to 0.7 μm. Between the single measurements, the probe is moved in the x-direction by the linear axis. The displacement between two measurements determines the measurement point distance Δx relative to the wire. The sensor detects the upper edge of the probe. Consequently, the measurements consider only the z-direction. Assuming that the used probe is not rotationally symmetric, a different behavior in the y- and z-directions is expected and the angular orientation of the probe around its length axis (x-axis) could influence the measurement results. Consequently, two measurements are performed for each test configuration. First, a measurement with the same probe section and probe orientation is performed for both guides. Following this, the probe is rotated 90° around its length axis (x-axis) and the measurements are repeated. For each measurement, the probe is fed in fixed steps of distance Δx until reaching position $x = 10$ mm and measured for each step, before being pushed back to the starting position. This measurement procedure is repeated 10 times. The radial deviation e is calculated by relating each measurement to the initial value at the position of $x = 0$ mm. The centering force of the guides is assumed to be direction-independent. Hence, the orientation of the guides is considered negligible. Nevertheless, G1 is oriented during testing in a way so that the uppermost of the three springs is aligned with the z-axis.

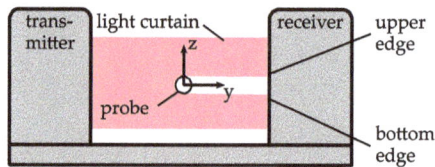

Figure 6. Measurement setup—radial deviation.

For a proper interpretation of the measurements, it is important to distinguish different influences. One influence is expected to be the contour accuracy of the probe, which is not necessarily rotationally symmetric. Therefore, the base material of the linked parts, which is hard-drawn wire (1.4301/AISI 304) with a diameter of $d_{wire} = 360$ μm with bright surface produced in bars, is used instead of the linked parts for the first experiments. The measurement point distance is set to $\Delta x = 1$ mm. Figures 7 and 8 illustrate the results. The range of the mean value from the minimum to the maximum is beneath 10 μm for the orientation of 0° and beneath 5 μm for the orientation of 90° for both guides. The standard deviation of the measurements is close to the sensor resolution, but obviously smaller for guide G2. Within the diagrams, the measurements show a similar curve shape for the same angular orientation but different guides. Comparing the curve shapes between the two diagrams respectively, a clear difference can be observed between the different angular orientations. This can be explained by the fact that even the raw wires show some contour deviations due to being not ideally straight and rotationally symmetric.

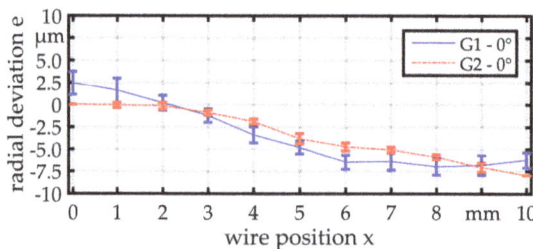

Figure 7. Comparison of radial deviation for G1 and G2 measured with wire, $d_{wire} = 360$ μm.

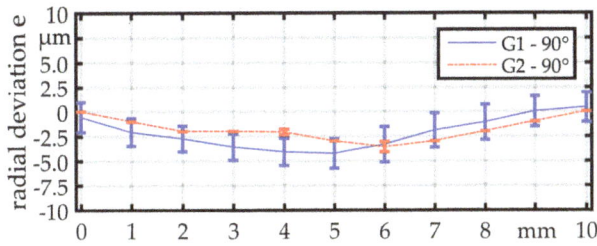

Figure 8. Comparison of radial deviation for G1 and G2 measured with wire, $d_{wire} = 360$ μm.

Finally, G1 and G2 are tested with real linked parts. For further identification of the influence of the contour of the parts, two different linked parts are used (Figure 9). Linked part 1 shows a good rotational symmetry and no measurable offset $o_{p1} \approx 0$ μm between the two wires at both sides of the part. Linked part 2, in contrast, shows an offset o_{p2}. The frame diameter of $d_{frame} = 360$ μm and the part diameter of $d_{part} = 660$ μm is equal in both cases. The length of the parts is $l_{part\,1} \approx 1200$ μm and $l_{part\,2} \approx 1000$ μm. The measurement procedure is the same as before, but the measuring point distance is reduced to $\Delta x = 0.2$ mm. Furthermore, the measurement is again performed two times for each linked part and, for the second measurement, the linked part is rotated 90° around its length axis (*x*-axis). The parts are illustrated in Figure 9 in both orientations. The measured offset of linked part 2 is $o_{p2,0°} = 165$ μm for the orientation of 0° and $o_{p2,90°} = 225$ μm for the orientation of 90°.

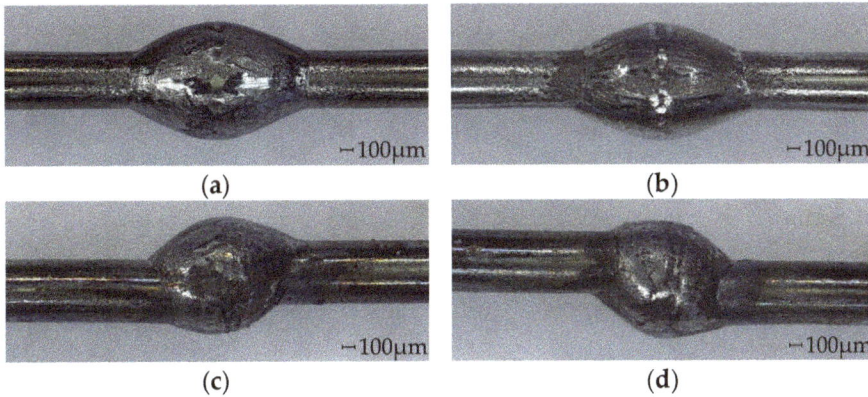

Figure 9. (**a**) Linked part 1 at 0°, (**b**) Linked part 1 at 90°, (**c**) Linked part 2 at 0°, and (**d**) Linked part 2 at 90°.

In the measurements with G1 (Figure 10a), the regions where the bearing balls are in contact with the parts can clearly be recognized by the unsteady curve. The range of the deviation is about 40 μm in the zones where only the wire is in contact with the bearing balls. The overall range of the deviation is about 160 μm. A similar behavior can be observed in the measurements for G2 (Figure 10b), but the curve is smoother. The overall range of the deviation is about 50 μm. Before the part enters the guide (*x* = 1 mm) and after it has passed through (*x* = 9 mm), the deviation e is close to 0 μm. The curves for the two different angular orientations are quite similar. As expected, the deviation for linked part 2 is much higher and in a range of 225 μm (Figure 11). The deviation is on a constant level, when the part is not within the guide, but there is a significant offset between the two described regions when the part is before and when it is behind the guide. For both guides, the curves are similar, but for G2, the curve is smoother and shows a smaller standard deviation.

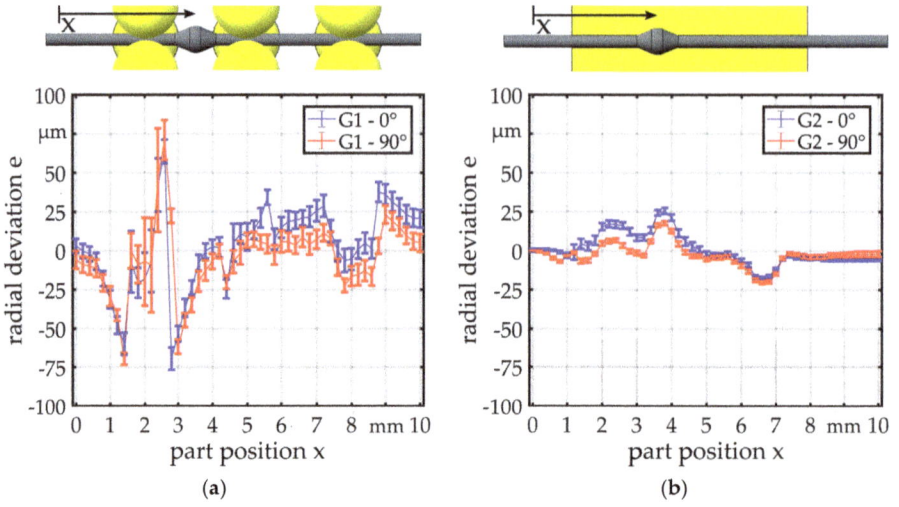

Figure 10. Radial deviation for (**a**) G1 and (**b**) G2 measured with linked part 1 (no offset between wire ends).

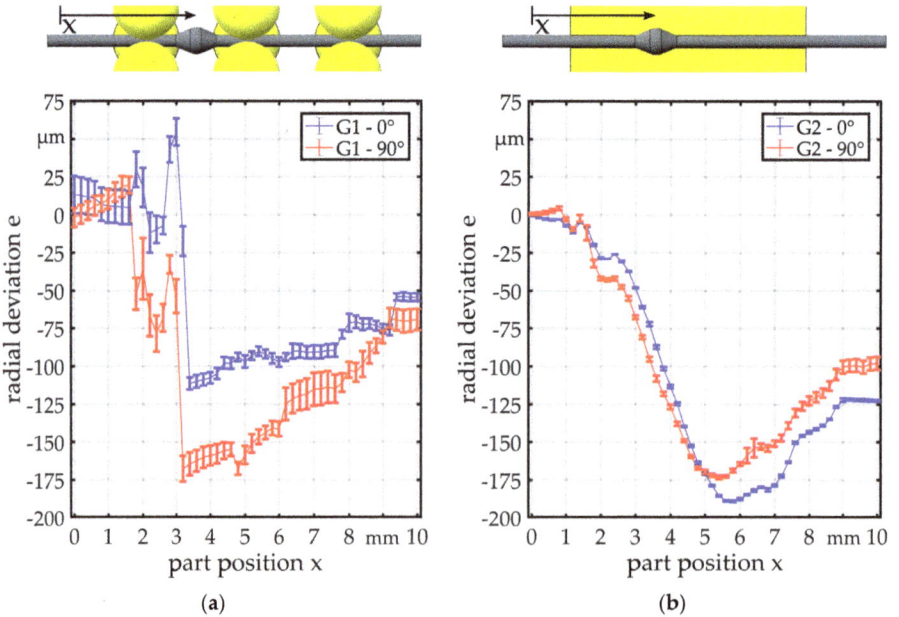

Figure 11. Radial deviation for (**a**) G1 and (**b**) G2 measured with linked part 2 (offset between wire ends).

4.2. Vibrations

For the vibration analysis, an application-specific testing is performed, which allows a qualitative comparison of the different guide concepts under the aspect of suppression of radial vibrations (Figure 12).

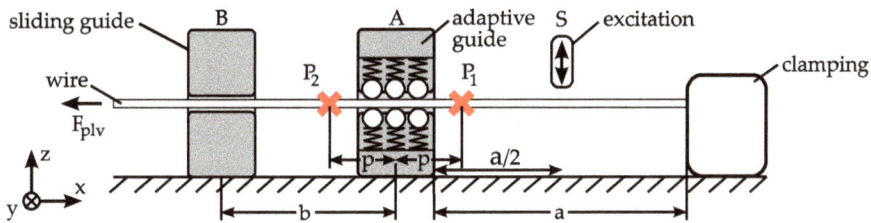

Figure 12. Setup vibration analysis, a = 230 mm, b = 60 mm, p = 17 mm.

All measurements are performed with a raw wire with a diameter of d_{wire} = 360 μm. The wire is clamped at one end, fitted within a sliding guide (position B) and preloaded with F_{plv} = 25 N at the other end. The different guides are mounted in position A during testing. At point S, the wire is excited by displacing and releasing it in a repeatable way, so that it vibrates in its eigenfrequencies. The wire is simultaneously measured at the points P_1 and P_2. The same sensor in previous measurements is used and the initial value is set to zero in order to calculate the radial deviation e. For comparison, a standard non-adaptive sliding bushing with a diameter of d_{sg} = 800 μm is used. Results are shown in Figure 13. The bushing provides a clearance of 0.22 mm between the wire and guide, as may be necessary in the worst case for a linked part. The wire can almost vibrate freely on both sides of the guide. The amplitude is about 8 μm. The results for the guides G1 and G2 are illustrated in Figures 14 and 15, respectively. Both show a good suppression of radial vibration for the wire segment behind the guide. Considering the vibrations at point P_1, G2 shows a better damping, as could be expected for an elastomer. In the measurements, no vibration can be seen. There is only a variation in the range of the sensor resolution.

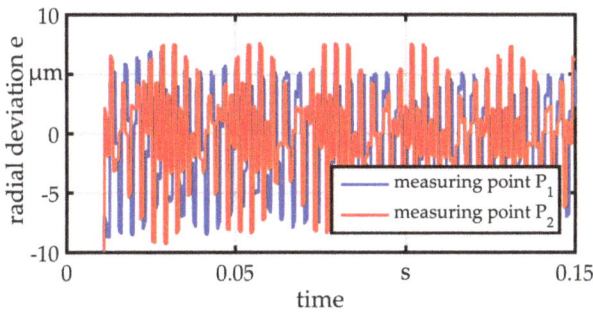

Figure 13. Vibration measurement sliding guide d_{sg} = 0.8 mm.

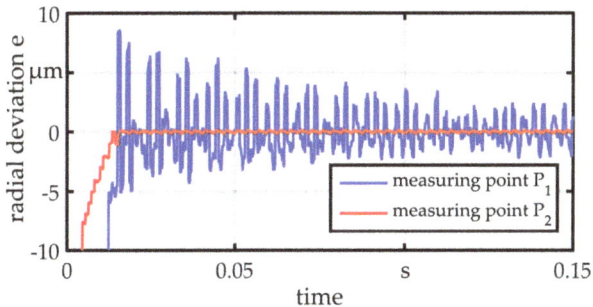

Figure 14. Vibration measurement guide G1.

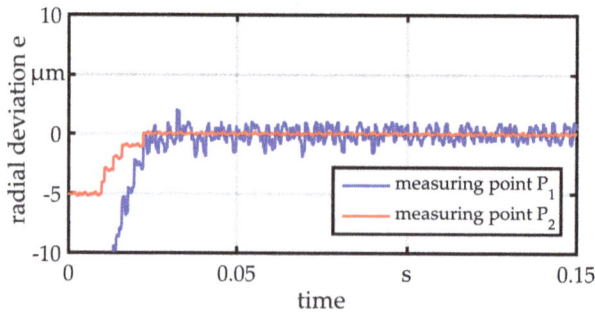

Figure 15. Vibration measurement guide G2.

4.3. Feed Force

The developed passive principle is based on the idea of using the feed force for a diameter adaption of the guide. When this adaption is realized by spring elements with a constant spring rate, the feed force changes during feeding. Measurements are performed with a force sensor (Figure 16). The feed distance is 20 mm, so that the part is able to completely pass through the guide.

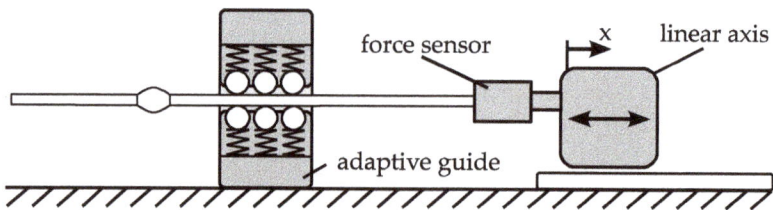

Figure 16. Measurement setup—feed force.

Figures 17 and 18 illustrate two exemplary results at a feed velocity of v = 10 mm/s. In both measurements, the wire is tensioned at the start of the motion and the force rises quickly. Only the frame moves through the guide and the force is relatively constant with $F_{G1,frame}$ = 0.02 N and $F_{G2,frame}$ = 0.03 N. In the measurements of G1, the force peaks of $F_{G1,peak}$ = 0.06 N appear when the part comes in contact with the bearing balls. The distance $m \approx 3$ mm between the balls can be recognized in the measurement. For G2, the force rises up to $F_{G2,peak}$ = 0.1 N when the part passes through the guide. In this case, there is a little difference between the zone of elevated force and the length of the guide, which is l = 8 mm. This can be explained by the lower axial stiffness of the elastomer element.

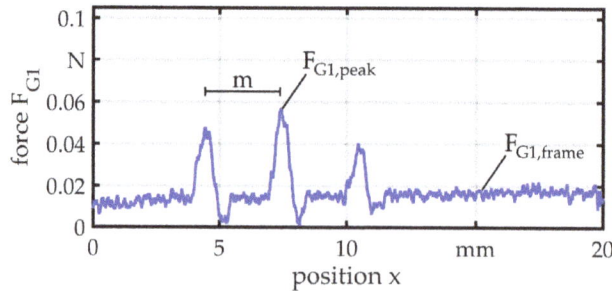

Figure 17. Force measurement—G1.

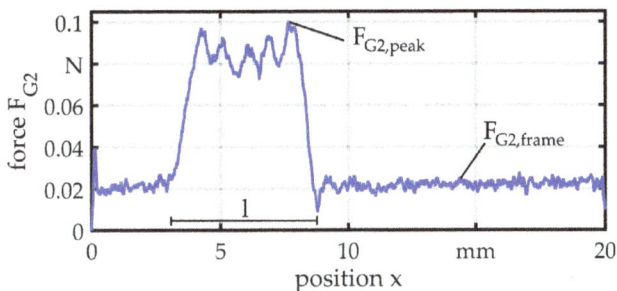

Figure 18. Force measurement—G2.

5. Conclusions

A concept for the passive and adaptive guiding of linked parts has been developed. Based on the concept, two prototypes were built and examined under the aspects of guiding precision, suppression, and damping of radial vibrations as well as influence on the feed force.

It is shown that adaptive guides improve the guiding precision compared to non-adaptive guides, but the precision is still influenced by the contour of the parts that are guided. This influence is more or less distinct depending on the guiding concept. Regarding the feed force, an influence of the adaption is a direct consequence of the passive nature of the guiding. The feed force is used for the radial adaption and, consequently, the adaptive guides cause force peaks when parts pass through. These peaks can be clearly observed. They are low in comparison to the forces that are delivered by typically applied feed devices. Nevertheless, a reduction by further optimization of the kinematics is aspired to decrease the amplitude of these peaks. Otherwise, these force peaks could induce vibrations to the system and decrease the accuracy. In the vibration measurements, it is shown that both guide concepts are generally suited to suppress the transmission of vibrations between different sections of the linked parts. Concerning the damping of vibrations close to a process guide, G2 shows a better performance. Based on the developed function scheme, further concepts can be realized to further improve the damping behavior.

Supplementary Materials: The following are available online at www.mdpi.com/2072-666X/8/6/169/s1.

Acknowledgments: The authors gratefully acknowledge the financial support by Deutsche Forschungsgemeinschaft (DFG, German Research Foundation) for Subproject C5 "Linked Parts" within the SFB 747 (Collaborative Research Center) "Micro Cold Forming—Processes, Characterization, Optimization". Furthermore, we thank Subproject A3 "Material Accumulation" for the production of the linked parts containing the preforms.

Author Contributions: P.W. conceived and designed the experiments; P.W. performed the experiments; P.W. analyzed the data; B.K. and C.S. provided advice on experiment design and data analysis; B.K., C.S. and P.W. wrote the paper. All authors have read and approved the final manuscript.

Conflicts of Interest: The authors declare no conflict of interest.

References

1. Vollertsen, F. Categories of size effects. *Prod. Eng. Res. Dev.* **2008**, *2*, 377–383. [CrossRef]
2. Van Brussel, H.; Peirs, J.; Reynaerts, D.; Delchambre, A.; Reinhardt, G.; Roth, N.; Weck, M.; Zussman, E. Assembly of microsystems. *CIRP Ann. Manuf. Technol.* **2000**, *49*, 451–472. [CrossRef]
3. Fantoni, G.; Santochi, M.; Dini, G.; Tracht, K.; Scholz-Reiter, B.; Fleischer, J.; Lien, T.K.; Seliger, G.; Reinhart, G.; Franke, J.; et al. Grasping devices and methods in automated production processes. *CIRP Ann. Manuf. Technol.* **2014**, *63*, 679–701. [CrossRef]
4. Tichem, M.; Lang, D.; Karpuschewski, B. A classification scheme for quantitative analysis of micro-grip principles. *Assem. Autom.* **2004**, *24*, 88–93. [CrossRef]

5. Fantoni, G.; Porta, M. A critical review of releasing strategies in micro parts handling. *Int. Fed. Inf. Process.* **2008**, *260*, 223–234.

6. Okazaki, Y.; Mishima, N.; Ashida, K. Microfactory—Concept, history, and developments. *J. Manuf. Sci. Eng.* **2004**, *126*, 837–844. [CrossRef]

7. Das, A.; Murthy, R.; Popa, D.; Stephanou, H. A multiscale assembly and packaging system for manufacturing of complex micro-nano devices. *IEEE Trans. Autom. Sci. Eng.* **2012**, *9*, 160–170. [CrossRef]

8. Arentoft, M.; Eriksen, R.S.; Hansen, H.N.; Paldan, N.A. Towards the first generation micro bulk forming system. *CIRP Ann. Manuf. Technol.* **2011**, *60*, 335–338. [CrossRef]

9. Mahshid, R.; Hansen, H.N.; Arentoft, M. Characterization of precision of a handling system in high performance transfer press for micro forming. *CIRP Ann. Manuf. Technol.* **2014**, *63*, 497–500. [CrossRef]

10. Fleischer, J.; Herder, S.; Leberle, U. Automated supply of micro parts based on the micro slide conveying principle. *CIRP Ann. Manuf. Technol.* **2011**, *60*, 13–16. [CrossRef]

11. Kuhfuss, B.; Moumi, E.; Tracht, K.; Weikert, F.; Vollertsen, F.; Stephen, A. Process chains in microforming technology using scaling effects. In Proceedings of the ESAFORM 24, Belfast, UK, 27–29 April 2011; pp. 535–540.

12. Fu, M.W.; Chan, W.L. A review on the state-of-the-art microforming technologies. *Int. J. Adv. Manuf. Technol.* **2013**, *67*, 2411–2437. [CrossRef]

13. Razali, A.; Qin, Y. A review on micro-manufacturing, micro-forming and their key issues. *Proceedia Eng.* **2013**, *53*, 665–672. [CrossRef]

14. Ghassemali, E.; Tan, M.-J.; Jarfors, A.; Lim, S.C.V. Progressive microforming process: towards the mass production of micro-parts using sheet metal. *Int. J. Adv. Manuf. Technol.* **2012**, *66*, 611–621. [CrossRef]

15. Merklein, M.; Stellin, T.; Engel, U. Experimental study of a full forward extrusion process from metal strip. *Key Eng. Mater.* **2012**, *504*, 587–592. [CrossRef]

16. Bruenning, H. Thermal free form heading. In *Micro Metal Forming*; Vollertsen, F., Ed.; Springer: Berlin/Heidelberg, Germany, 2013; pp. 188–199.

17. Kuhfuss, B.; Moumi, E.; Piwek, V. Micro rotary swaging: Process limitations and attempts to their extension. *Microsyst. Technol.* **2008**, *14*, 1995–2000. [CrossRef]

18. Wilhelmi, P.; Moumi, E.; Schenck, C.; Kuhfuss, B. Werkstofffluss beim Mikrorundkneten im Linienverbund. In Proceedings of the 5. Kolloquium Mikroproduktion, Aachen, Germany, 16–17 November 2015.

19. Wilhelmi, P.; Schenck, C.; Kuhfuss, B. Linked micro parts referencing system. *J. Mech. Eng. Autom.* **2017**, *7*, 44–49.

20. Hansen, H.N.; Carneiro, K.; Haitjema, H.; De Chiffre, L. Dimensional micro and nano metrology. *CIRP Ann. Manuf. Technol.* **2006**, *55*, 721–743. [CrossRef]

21. Wilhelmi, P.; Schenck, C.; Kuhfuss, B. Diameter adaptive guides for wire-based linked micro parts. In Proceedings of the 4M/IWMF 2016 Conference, Kongens Lyngby, Denmark, 13–15 September 2016.

22. Isomura, K.; Tanaka, S.; Togo, S.; Esashi, M. Development of high-speed micro-gas bearings for three-dimensional micro-turbo machines. *J. Micromech. Microeng.* **2005**, *15*, 222–227. [CrossRef]

23. Zhang, Q.; Shan, X. Dynamic characteristics of micro air bearings for microsystems. *Microsyst. Technol.* **2008**, *14*, 229–234. [CrossRef]

24. Fernandez, V.; Reyne, G.; Cugat, O.; Gilles, P.; Delamare, J. Design and modelling of permanent magnet micro-bearings. *IEEE Trans. Magn.* **1998**, *34*, 3592–3595. [CrossRef]

25. Coombs, T.; Samad, I.; Ruiz-Alonso, D.; Tadinada, K. Superconducting micro-bearings. *IEEE Trans. Appl. Supercond.* **2005**, *15*, 2312–2315. [CrossRef]

26. Waits, C.; Geil, B.; Ghodssi, R. Encapsulated ball bearings for rotary micro machines. *J. Micromech. Microeng.* **2007**, *17*, 224–229. [CrossRef]

27. Tian, Y.; Shirinzadeh, B.; Zhang, D. Design and dynamics of a 3-DOF flexure-based parallel mechanism for micro/nano manipulation. *Microelectron. Eng.* **2010**, *87*, 230–241. [CrossRef]

28. Vekteris, V.J. Principles of design and classification of adaptive bearings. *Tribol. Trans.* **1993**, *36*, 225–230. [CrossRef]

MDPI AG

St. Alban-Anlage 66

4052 Basel, Switzerland

Tel. +41 61 683 77 34

Fax +41 61 302 89 18

http://www.mdpi.com

Micromachines Editorial Office

E-mail: micromachines@mdpi.com

http://www.mdpi.com/journal/micromachines

www.ingramcontent.com/pod-product-compliance
Lightning Source LLC
Chambersburg PA
CBHW041214220326
41597CB00033BA/5896